"This handbook should be required reading not only for technicians, but for any professionals who expect to have their articles published in technical trade magazines."

John R. Gyorki
Senior Editor, Machine Design Magazine

"You've written in a style that's easily understood by anyone. Very simply, for prospective technical writers, your book is an absolute MUST!"

Harold Ort N2RLL
Editor, RADIO magazine

"Technical Writing for Technicians provides an excellent introduction to the field of technical writing. This one-of-a-kind book should help expand the career opportunities of many technicians."

Teri Scaduto
Popular Electronics Magazine

"This book fills a need for a quick writing reference for technical people."

John L. Fike Ph.D. P.E.
Associate Professor of Electronics and Telecommunication Engineering Technology; Director of the Center for Telecommunication Technology Management; Victor H. Thompson Professor at Texas A&M University

"I think your coverage of work responsibilities and attention to detail is very good. In terms of teamwork, you describe an exemplary worker. You give the reader a good picture of what management wants."

Robert Schetgen KU7G
Editor, ARRL Handbook

TECHNICAL WRITING
for
TECHNICIANS

Warren R. Freeman

Contemax Publishers

Minneapolis, Minnesota

TECHNICAL WRITING for TECHNICIANS

TECHNICAL WRITING
for
TECHNICIANS

by Warren R. Freeman

Published by: CONTEMAX PUBLISHERS
17815 24th Avenue North
Minneapolis, Minnesota 55447
U.S.A.
SAN 298-4970

Copyright ©1995 Warren R. Freeman
Minneapolis, Minnesota
Printed in the United States of America

All rights reserved. No part of this book may be reproduced or transmitted in any form or by any means, electronic, or mechanical, including photocopying, recording, or by any information storage or retrieval system without written permission from the author, except for brief quotations in a review.

Library of Congress Catalog Card Number: 94-90876

Freeman, Warren R.
Technical Writing for Technicians

ISBN 0-9644739-0-9

TECHNICAL WRITING for TECHNICIANS

To my wife, Karin

TECHNICAL WRITING for TECHNICIANS

Disclaimer:
While this book explains certain procedures used to prepare a useful technical manual to accompany hardware devices, knowledge of this information cannot guarantee the reader a position as a technical writer.

TECHNICAL WRITING for TECHNICIANS

Contents

Section		Page
I	**Introduction**	**1-1**
	The Hardware Technical Writer	1-1
	Work Relationships	1-2
	Possible Employers	1-2
	The Hardware Technical Manual	1-5
	Overview of a Technical Manual	1-6
	The Appearance of a Technical Manual	1-6
	Writer's Equipment	1-8
	Using a Computer	1-8
II	**Preparation for Writing**	**2-1**
	Introduction	2-1
	Kinds of Manuals	2-1
	Stages of Manuals	2-1
	Hardware	2-2
	The Subject Hardware	2-2
	Hardware Availability	2-2
	Engineering Documentation	2-3
	Important Considerations	2-6
	Who is My Audience?	2-6
	The Due Date and Writer Hours	2-6
	Questions to Ask	2-7
	Writing Start	2-7
	The Estimate	2-7
	Section or Chapter?	2-9
	Planning Ahead	2-9
	Support	2-10
	Outlining	2-12
III	**Writing Style**	**3-1**
	How to Write Clearly	3-1
	Style	3-1
	Panel Nomenclature	3-2
	Punctuation	3-2
	Sentence Structure	3-5

TECHNICAL WRITING for TECHNICIANS

Contents (Cont)

Section		Page
IV	**Text Preparation**	**4-1**
	General	4-1
	Draft	4-1
	The Starting Point	4-1
	Illustrations	4-3
	Tables	4-6
	Warnings and Cautions	4-6
	Record Keeping	4-7
	Estimating	4-8
	The Page Unit	4-8
	Draft to Formatted Document	4-8
	A Computer in the Hardware	4-9
V	**Illustrations**	**5-1**
	Illustrations	5-1
	Discussion	5-1
	Types of Drawings	5-1
	Line Art	5-2
	Introduction	5-2
	Sketching	5-2
	Drawing	5-3
	Planning	5-10
	Checking Artwork	5-11
	Sizing Line Art	5-11
	Type Size and Typeface	5-12
	Photographs	5-13
VI	**Reviews**	**6-1**
	Introduction	6-1
	In-House Reviews	6-1
	Military Reviews	6-2
VII	**Layout and Printing**	**7-1**
	Layout	7-1
	The Layout Expert	7-1
	Proofreading	7-1
	Printing	7-2
VIII	**Writing a Sample Manual**	**8-1**

Note: Section VIII has its own table of contents. Refer to Page 8-9.

TECHNICAL WRITING for TECHNICIANS

List of Illustrations

Figure	Title	Page
1-1	The Writer's Position on the Team	1-3
1-2	Writer's Equipment	1-9
2-1	Controls and Indicators	2-5
2-2	Suggested Section Headings	2-11
2-3	Example of a Detailed Outline (partial)	2-14
4-1	Placing a Large Figure on Opposing Pages	4-4
4-2	Table Example	4-8
5-1	Orthographic and Isometric Examples	5-5
5-2	Code of Lines	5-6
5-3	An Orthographic Drawing	5-7
5-4	An Isometric Drawing	5-8
5-5	A Circle in Isometric	5-9

Appendix A	**General Information**	**A-1**
Resistor Color Code		A-1
American Wire Gauge to Square Millimeters		A-2
Some Constants		A-3
Fahrenheit and Centigrade Conversions		A-4
Conversion Factors in Alphabetical Order		A-7
Some Abbreviations		A-14
Greek Alphabet		A-15
Mathematical Notation		A-16
Proofreader's Marks		A-17
Sinking and Sourcing Conventions		A-18

Appendix B	**References**	**B-1**

Index

Important note: This book is arranged on the lines of a technical manual used by the military, in which page, figure, and table numbering are by section. As an example: page 3-2 is the second page of Section 3, and Figure 5-3 is the third figure in Section 5.

TECHNICAL WRITING for TECHNICIANS

TECHNICAL WRITING for TECHNICIANS

Acknowledgements

I wish to gratefully acknowledge the help received from the following reviewers, for time taken from their families to read and provide most useful suggestions and comments on the manuscript of this book.

Technical content was reviewed by:
Robert Caneen, Robert J. Daly, and **Gary G. Good.**
General layout, English, and content were reviewed by
Mary S. Hendrix and **Mark R. Freeman.**
Constant support was given me by my wife, **Karin.**

My thanks to **Tandy Radio Shack** for their generous permission to use their equipment as the basis for the sample manual of Section VIII.

In writing this book, I have made every effort to present an accurate picture of the writer's work. I take full responsibility for any errors which may be found.

TECHNICAL WRITING for TECHNICIANS

> If I supply you a thought, you may remember it and you may not. But if I can make you think a thought for yourself, I have indeed added to your stature.
>
> Elbert Hubbard

TECHNICAL WRITING for TECHNICIANS

PREFACE

In my experience as a technical writer, I have found that those who become the best hardware technical writers usually have a technical background, and a couple of years of study beyond high school, which included English and physics.

In working for different companies, I was often approached by technicians who wanted to know about technical writing. Most of them could do a commendable job as technical writers, if given a little encouragement to make the jump. For those with such hands-on experience, this book has been written.

Of secondary interest is my desire to inform managers about technical writing and its importance, especially to marketing. A well-written manual is an effective sales tool, and ensures customer satisfaction after the subject hardware is delivered.

The information given in this book is general. It is a guide, subject to bending to the immediate needs of an employer or a specific situation.

the author

TECHNICAL WRITING for TECHNICIANS

The author validating his manual

SECTION I
INTRODUCTION

This section introduces the reader to technical writing, describes some choices of writer employment, and explains the structure of a technical manual.

THE HARDWARE TECHNICAL WRITER

The hardware technical manual is written by one or more technical writers. The ideal technical writer would be the engineer who designed the equipment, who also has an unusual gift for clear writing. However, most engineers prefer to "engineer," and are reluctant to write.

This provides an interesting employment opportunity for a technician with a good command of English, who can organize technical material, and work well with technical people.

If you are trained as a technician, and have these abilities, consider becoming a technical writer. Technical writing is a satisfying occupation in which you are responsible for simplifying and explaining complex systems to the reader.

What are the advantages? Technical writing is a necessary and respected endeavor. In many companies, the technical writer is classified as Professional & Technical, a white-collar, non-union occupation.

A technical writer's income is somewhat higher than that of an average technician, but lower than that of an engineer. Work is normally performed in an office atmosphere, where there is

TECHNICAL WRITING for TECHNICIANS

always something new to learn. Pressure may build to get your manuals "out the door", but you will sense this in advance, and have some degree of control over your schedule. Sometimes you will have several manuals on your desk, and must determine the relative priorities based on actual need, availability of information, and supporting services.

A part of writing includes an ability to visualize and sketch fairly complex items, and the ability to use a personal computer. These skills can be developed either on the job, or through short evening classes.

You may ask, "What related kinds of work exist for a hardware technical writer"? The writer can prepare test specifications, edit contract reports, prepare advertising brochures, and if sufficiently creative, move into advertising or, if so inclined, into sales engineering, or management.

WORK RELATIONSHIPS. Figure 1-1 shows the basic interrelationship necessary to produce a manual. The writer works closely with the design and manufacturing engineering staff, using much of their design information to produce a technical manual which will support their hardware as it goes to the customer.

POSSIBLE EMPLOYERS. A large company may have a complete technical writing department with a manager, work directors, (lead writers), a staff of writers for both hardware and software, technical illustrators, Illustrated Parts Breakdown (IPB) writers (catalogers), a print shop and established working rules. In smaller companies, manual preparation may be the responsibility of the head of the sales, marketing, advertising, engineering, or field service department. If you

Introduction

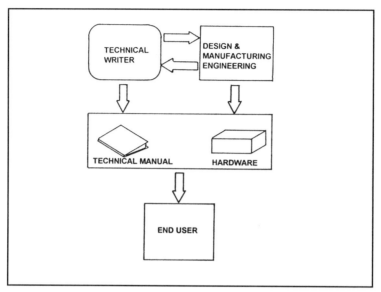

Figure 1-1. The Writer's Position on the Team

work in a small company, you may be the only writer, responsible for developing entire manuals, and may be called on to help prepare advertising materials.

Direct Hire. A larger company may elect to hire a writer directly, relying on the Human Resources (employment or personnel) department to screen for acceptable applicants. If hired directly, the writer receives full company benefits and is assigned to the writing department. From here, he (or she) may be given training, and most valuable, an opportunity to develop writing skills on actual writing assignments.

If the company is small, the writer may be directly employed by a company officer, and may receive minimal training... only experience.

TECHNICAL WRITING for TECHNICIANS

As to the path of progression, the junior writer may move up to writer in about 2 years, then to senior writer after about 2 to 5 more years. As of this printing, earnings, based on a 40-hour week average $20/hour for junior writers to about $35/hour and up for senior writers, depending on the industry. Note that company benefits must be considered a plus.

Before accepting any position, be sure to meet the person you will actually be working for, and determine through conversation whether you are relatively compatible. Ask specifically what you will be required to do (job description); about opportunities to advance; and details about the overall organization. For example, is the writing staff under marketing, engineering, field service, or directly under the president?

Indirect Technical Writers (Contractors). Frequently, a company will bring in a contract writer to fill their immediate needs. The writer may be hired by the company through an outside contractor (job shop), or may be hired on his/her own contract. If working through a job shop, the writer is the employee of the job shop, and may in time be eligible for their benefits. If on one's own contract, the writer will not receive company benefits.

What are the tradeoffs? If the work comes to an end, the job shop, through an extensive network of contacts can possibly find you a position more rapidly than you can on your own. They bill the company for your hourly wage plus their fee which may range from 10% to 100% of what you receive. With such a high monthly bill, the client company is most anxious that you apply yourself to the writing task immediately.

Introduction

If you are on your own contract, you can possibly ask for a slightly higher wage since the employer doesn't pay a fee, but then you must find the job. In contract writing, one must be aware that the job exists at the pleasure of the client. Some writers work as contractors for a short period of a few weeks to perform an assigned writing task, then go on to another company. Some are hired as direct employees by the client company. Others remain on contract for years.

In a given geographic area, acceptance of direct employment may limit a writer's opportunity to later move into contract writing in the same area because of non-competition agreements between the employer and local contract houses.

If you are working through a job shop and sense that the work is running out, immediately notify the job shop so they can find other employment for you. You may even find yourself holding down two jobs at one time. If you are on your own contract, you may be able to take time off until a new writing task becomes available, or you may want to find another company if work is not forthcoming. As a matter of honesty, when you start a job, see it through to completion if at all possible, and don't ever suggest a pay raise in the midst of a job.

THE HARDWARE TECHNICAL MANUAL

The hardware technical manual is a guidebook that accompanies equipment to the end user. The end user is always an individual, although the equipment may be purchased or leased by a company.

TECHNICAL WRITING for TECHNICIANS

OVERVIEW OF A TECHNICAL MANUAL. The purpose of the technical manual is to inform the user about the equipment, how to unpack it, install it, operate it, detect and correct malfunctions, and order replacement parts.

The subject equipment may be mechanical, electrical, electromagnetic, electronic, optical, or sonic. Its design may incorporate any combination of these and other disciplines.

With the increasing complexity of equipment, the hardware technical manual is often considered part of the deliverable equipment, and may actually be packaged in the same shipping crate or box with the equipment. With such visibility, the manual must be accurate, easy to read, and ready at the time of hardware shipment.

If shipment is made to a foreign user, the manual may be translated into the user's language. Manuals intended for translation must be carefully written using words which cannot be misinterpreted in translation.

A typical manual contains only facts, no advertising material, and generally describes the operation of only one piece of equipment.

THE APPEARANCE OF A TYPICAL MANUAL. The hardware technical manual, when printed and bound, usually consists of backprinted pages on which text and illustrations are artfully arranged, likely with text in two columns. The start of each line of text is aligned at a fixed distance from the left margin (left justified), and the line of text is adjusted

Introduction

to a fixed length to provide an even right edge to the column (right justified). If the text is in single column, the left margin will be justified, and the right margin may be justified. If text is not right justified, it is known as "ragged right".

A space across the top of each page may be reserved for the manual's title and section title. This is the header. A footer, a space across the bottom of the page, may carry the page number. Following the cover are the pages of front matter, consisting of the title page, copyright notice, table of contents, list of illustrations, and list of tables. Some manuals include a summary of warnings and cautions used throughout the manual.

The first page of text is the title page, always a right-hand page. This is page one, but may not bear a page number. The following text pages are numbered sequentially in Arabic numerals, or may be numbered to the end of each section using a hyphenated designation which identifies the section followed by the page number, for example: 1-1, 1-2, 2-1, 2-2.

The paper on which a manual is printed may be determined by the customer, or company considerations. The paper stock selected must be able to accept backprinting without show-through and of a quality to reproduce sharp halftone figures (photographs) if they are used.

The type of binding can consist of three-ring binders with the pages drilled to fit (loose leaf); comb binding, saddle-stitching using staples; or perfect binding, in which adhesive joins the cover and pages as seen in the common phone book and other "soft-cover" books.

TECHNICAL WRITING for TECHNICIANS

WRITER'S EQUIPMENT

In your workplace, you likely will be provided with desk space and a personal computer. Access to an office copier is most helpful. Personal equipment which you will need is minimal, as shown in Figure 1-2. All these materials can be carried in a briefcase.

As work progresses, a writer should assemble a basic library consisting at least of an English dictionary, a style guide, a dictionary of terms specific to the work, together with math and physics books for fast reference. Remember, these items are investments which will enable you to work more efficiently. In this connection, it is good practice to save receipts for all items of relatively high value so you can remove them from the company premises should the need arise.

USING A COMPUTER

The personal computer (PC) provides the writer with a more versatile and forgiving typewriter. Errors can be corrected on the screen, and parts of the text can be easily moved around. With a scanner, you can enter artwork into the PC where it can be sized and placed with the text. Your manual will be in a file of its own, and have a name by which you can find it. It can be printed out in part or in its entirety. Other manuals may reside in different files on the hard drive of your computer.

If another manual on your PC contains text that you can use, you can open its file alongside your open file and transfer a copy of the text of interest to your own file. As you work on your draft, you must be sure to "save" your text to the hard disc at regular intervals (manually every 10-minutes unless you have an autosave feature in operation) to avoid losing valuable material in the event of a power outage.

Introduction

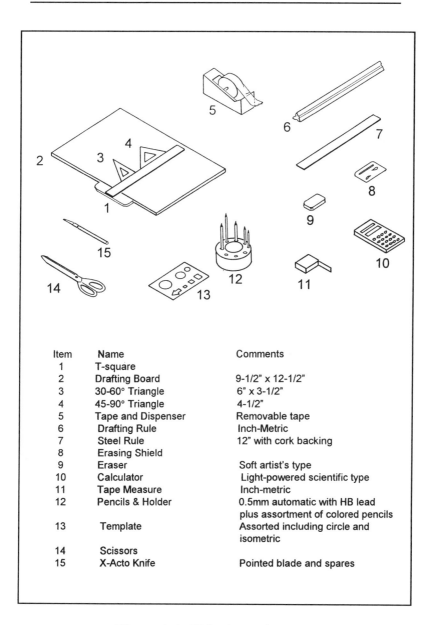

Item	Name	Comments
1	T-square	
2	Drafting Board	9-1/2" x 12-1/2"
3	30-60° Triangle	6" x 3-1/2"
4	45-90° Triangle	4-1/2"
5	Tape and Dispenser	Removable tape
6	Drafting Rule	Inch-Metric
7	Steel Rule	12" with cork backing
8	Erasing Shield	
9	Eraser	Soft artist's type
10	Calculator	Light-powered scientific type
11	Tape Measure	Inch-metric
12	Pencils & Holder	0.5mm automatic with HB lead plus assortment of colored pencils
13	Template	Assorted including circle and isometric
14	Scissors	
15	X-Acto Knife	Pointed blade and spares

Figure 1-2. Writer's Equipment

TECHNICAL WRITING for TECHNICIANS

When leaving your file, remember to save to the hard disc. The file should also be saved periodically to a "floppy" disc or "diskette". The most popular floppy disc measures 3-1/2 inches on a side, and is 1/8 inch thick, a wafer you can store easily, yet it can hold words and illustrations for a complete manual. A floppy must be protected from magnetic fields as well as mechanical damage.

It is a simple task to make things complex, but a complex task to make them simple.

                                                         Meyer's Law

SECTION II
PREPARATION FOR WRITING

INTRODUCTION

KINDS OF MANUALS. Common kinds of hardware manuals are: Installation, Operation, and Overhaul. An associated type, the Illustrated Parts Breakdown (IPB), written by catalogers, is not described here, but if available and current, may be of considerable value as it illustrates system parts, and lists part names and part numbers. Use an IPB with care. It is supposed to show parts in disassembly order, but may omit details which you as a writer must detect and explain to the user.

Installation and operation manuals are generally shipped with the equipment. They may be separate documents, or may be combined in a single manual. Overhaul manuals contain in-depth information for use by a repair department.

STAGES OF MANUALS. Manuals are prepared first as a manuscript with double-spaced text and available figures, some hand-drawn. Following approvals, they are converted to final form, ready to print.

In military manual preparation, the manual is termed "*draft preliminary*" with double-spaced text and available illustrations. Following reviews, the manual becomes *preliminary*, in which the text is laid out in two columns with all finished art incorporated. The military manual is published first as *preliminary*. After the manual has been used in the field for a period of time, possibly months, the military then orders

TECHNICAL WRITING for TECHNICIANS

the preliminary manual to be converted to *formal* with the addition of corrections and changes gained from their experience.

HARDWARE

THE SUBJECT HARDWARE. As part of your writing assignment, you must know, **and have in writing,** the exact description of the subject hardware about which you will be writing. This means the exact name, model number, and part number. This vital information must come from someone with **direct responsibility** for your work, preferably the manager. In some cases there will be more than one version of the equipment, then you must know which versions are to be included in the manual. Learn all you can about the equipment. Find out what support equipment is required to make it work.

If support equipment is required, is it described in its own manual? If so, does the user have access to that manual? If not, you may have to describe it in some detail in your manual. In any event, you will have to show how to use it or connect it to the subject equipment.

Examples of support equipment are: power supplies, alignment equipment, and personal computers (PCs) including laptop computers. Identification of such information may be through engineering test or acceptance specifications, or by contact with the responsible project engineer. Up-to-date price sheets available through the marketing department may list related equipment to be sold with the subject hardware.

HARDWARE AVAILABILITY. If the hardware is available for you to look at, and handle, you are indeed fortunate.

Preparation for Writing

Even more so if it is the exact item, and not a prototype or an older model.

Be sure to ask how long it will be available to you, and act accordingly to learn all you can. If it is a production item, built in your company facilities, watch one being assembled. In any event, find out what the controls and indicators do, where the electrical components are, and the locations of connectors and user-replaceable parts such as fuses, and indicator lamps.

Make sketches, and take careful notes. Notes should include the color of indicator lamps. Photographs, if permitted, are helpful.

There is no substitute for "hands-on" operation of the hardware if you can gain permission to work with it. Figure 2-1 shows a control panel with an accompanying listing of the controls and indicators. Notice that by using a table, and numbering the items in the figure, the figure is not cluttered with names, and the table can include a short description of the function of each item.

ENGINEERING DOCUMENTATION

There will usually be a body of engineering documentation for the subject hardware. This may consist of a proposal, specifications for design, test and acceptance, drawings, parts lists, engineering change notices (ECN's, ECO's, or EO's) and materials used for presentations.

In a large company, specifications, drawings and engineering change notices (ECNs) will be available through a documen-

TECHNICAL WRITING for TECHNICIANS

tation center. In a smaller company, this information may come from the responsible engineer. You must review all such information for validity... <u>does it apply directly to your subject equipment</u>, or could it be useful from an informational standpoint? Copies of useful data must be organized and placed in your writing file as support for your writing effort.

When filing such material, it is helpful to write the date on it and also include its source. As a rule this support material, called a "data base" is kept for at least a year after your manual is printed, and may have value in follow-on efforts such as hardware updates, or manuals on similar equipment.

A high degree of organization is inherent in engineering. There are "top drawings" for the subject equipment which point out all major items. Each major item is itself the subject of a drawing, and each such drawing has its subordinate drawings.

Generally the top drawing will have a "tree" with part numbers and drawing numbers of subordinate drawings. An exception to this neat "tree" arrangement occurs when the subject device is assembled from "piece parts," parts that are available to engineering, all having unrelated part numbers and no top drawing. In such a case, you may need help from an engineer to relate the parts. In the engineering world, two or more piece parts which are fastened together are called an "assembly."

Be especially careful in collecting specifications, drawings, and parts lists. Obtain and work only with the **latest drawing** revisions, get copies of any engineering change

Preparation for Writing

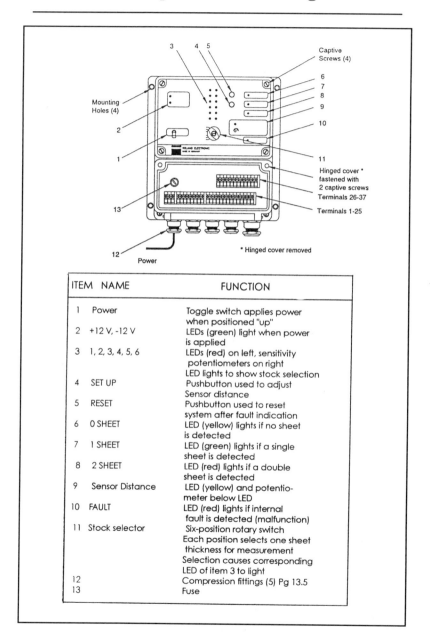

ITEM	NAME	FUNCTION
1	Power	Toggle switch applies power when positioned "up"
2	+12 V, -12 V	LEDs (green) light when power is applied
3	1, 2, 3, 4, 5, 6	LEDs (red) on left, sensitivity potentiometers on right LED lights to show stock selection
4	SET UP	Pushbutton used to adjust Sensor distance
5	RESET	Pushbutton used to reset system after fault indication
6	0 SHEET	LED (yellow) lights if no sheet is detected
7	1 SHEET	LED (green) lights if a single sheet is detected
8	2 SHEET	LED (red) lights if a double sheet is detected
9	Sensor Distance	LED (yellow) and potentiometer below LED
10	FAULT	LED (red) lights if internal fault is detected (malfunction)
11	Stock selector	Six-position rotary switch Each position selects one sheet thickness for measurement Selection causes corresponding LED of item 3 to light
12		Compression fittings (5) Pg 13.5
13		Fuse

Figure 2-1. Controls and Indicators

2-5

TECHNICAL WRITING for TECHNICIANS

notices (ECNs) which have been issued, or are pending against the drawing and its parts list. By talking with engineering personnel <u>directly</u> <u>concerned</u> with the subject hardware, you can learn of up-coming changes which may affect your work. Avoid basing your work on unreleased drawings, as they can change without the issuance of an ECN. Generally, the finished manual will be based on released specifications and drawings.

IMPORTANT CONSIDERATIONS

WHO IS MY AUDIENCE? Before starting to write, it is necessary to know who will use your new manual, and the depth of experience he or she has had with the subject hardware or similar equipment. This information will guide your writing presentation, and possibly your choice of words. This is especially important if the manual is to be translated. Your manager or the responsible marketing manager should be able to advise you. If you are writing for the military, the contract may specify the grade (education) level of the user and may limit the writing to a certain level of word and sentence complexity. This complexity is checked by a formula requiring the counting of words per sentence and word length. Fortunately this task can be done by computer.

THE DUE DATE AND WRITER HOURS. The manual due date and an estimate of the writer hours must come from management. The due date must be clearly defined. What is due?... is it your manuscript... or the finished manual? At this time, you should be given the <u>estimated writer's hours in writing</u>, and the writing cut-off date. The cutoff or "freeze" date is the date beyond which no further changes or additions can be made to your manual. In practice, however, this date is

Preparation for Writing

frequently ignored to accomodate late changes to hardware, and may cause compression of your work schedule.

QUESTIONS TO ASK. Ask your manager the following questions:

- The name of the project engineer, contracting officer, field service representative, and anyone else who has direct responsibility for the subject hardware.
- How many reviews will the manual go through?
- At what stages in the course of writing will reviews be held? How long will they take?
- Is the review time and time required to make corrections deducted from the writing time?
- What other departmental services must be used, and how heavy are their workloads?
- What about page layout, artwork and photography? Who controls priorities?
- What will the "deliverable" consist of? printed and bound copies? "camera ready" copy with all finalized art and photographs, or a clean manuscript with your own hand-drawn, but accurate sketches?

WRITING START

THE ESTIMATE. Your first impulse might be to start writing immediately. Wait a minute! ... what information is available, and how firm is it?

How did the manager arrive at the writer-hours estimate which you received at the start of your writing assignment? This estimate was likely prepared at an earlier date by an experi-

TECHNICAL WRITING for TECHNICIANS

enced writer using information available from the contracts department or from marketing. It may have been accurate when written, but because of developments and changes, it may be out-of-date. Your manager may not be aware of recent engineering changes which may impact your writing.

You must now determine the truth for yourself by "pulling" a set of the latest specifications and drawings, and talking with engineers and contract administrators who are <u>directly responsible</u> for the subject hardware. Be advised that their time and yours are valuable. As a matter of courtesy, set up an appointment, then be on time. If you find a disparity between the estimate you received, and what you can see in the size of the manual, and the time you will need, discuss this at the earliest opportunity with your manager. <u>Do not delay and overrun your hours.</u>

If you find the hardware design is not yet completed, and drawings are not released, or release is delayed, it must be your manager's decision when you should start, and what part of the writing job you should first attack. Most likely, you will be advised to work only on the firm areas for which most drawings have been released. Avoid getting into a frustrating "tail-chase" where you are attempting to keep up with rapid engineering changes.

As a reminder, your department will have a dollar allocation for completion of the technical manual, and a writer's hours must be profitably spent in order to complete the job "within budget and on schedule."

Preparation for Writing

SECTION or CHAPTER? The hardware technical manual is generally organized by sections or chapters, with each section or chapter starting on a right-hand page and developing within its body a central concept which differs from, but supports the other sections or chapters. If your company is a sub-contractor for the subject hardware, direction for manual preparation may appear in the contract under 'manuals requirement' which may list specifications, or refer to a sample manual. Be sure to review such requirements with your department manager.

In the absence of such direction, your department may have its own 'writer's style guide', or you may be given existing manuals to be used as models for your manual. In the absence of direction, a suggested organization is shown in Figure 2-2.

PLANNING AHEAD. Take a few minutes to put yourself in the position of the user, to envision what you would expect from your manual.
You, the user, are notified that the subject equipment has arrived, packed in a large box. As you lift the top flaps of the box, you first see the manual wrapped in plastic. Below is the subject hardware. You remove the manual, unwrap and open it. At this point, you know the purpose of the hardware, but have had no experience with it. The pressure is on to get it "up and running" as soon as possible.

From the manual's Table of Contents you see:

INSTALLATION

"Ah ha... here's where to start!" "This will tell me all I'll need to know to get the hardware in place and wired up."

TECHNICAL WRITING for TECHNICIANS

" The next section, ALIGNMENT, will help me get the optics correctly adjusted." The OPERATION section will lead me step-by-step to get the most out of this equipment. I'm sure glad they wrote a manual that's so easy to follow."

Placing yourself in the user's position is helpful, and will put you, the writer, in the proper frame of mind to consider what a manual <u>ought</u> to include. However, your employer is in business, and intends to remain in business, so you can't "give the customer a Cadillac when he ordered a Ford."

There is a schedule to meet and a limitation on writing hours, and the ultimate size and cost of your manual. The big question you must answer is: How much, and what detail should the manual contain to cover the subject, but not overrun cost / schedule / page-count limitations?

SUPPORT. Which groups can I count on for support in preparing my manual?

In a large writing department, there will be a department manager, assisted by work directors, technical editors, editors, page layout specialists, technical illustrators, catalogers, provisioners, photographers, and possibly a reproduction services group.

To explain: The manager has overall responsibility for the department; the work director (lead writer) oversees the day-to-day work of writers, the technical editor reviews the draft manual by section and in its entirety for technical accuracy, the editor looks for correct English and for consistency in organization. Layout specialists use a PC or a mainframe computer to format the manuscript, fitting the text and art into

Preparation for Writing

```
Title Page
Table of Contents/ List of Illustrations/List of Tables

List of Applicable Warnings and Cautions
Section
   I       Description
               General
               Controls and Indicators
               Theory of Operation
   II      Installation
               Receiving & Inspection
               Planning the Installation
   III     Alignment
   IV      Operation
   V       Maintenance & Troubleshooting
   VI      Design Data
               Models and Part Numbers
               Specifications
               Illustrated Parts Breakdown
               Optional Equipment
           Glossary
           Index
```

Figure 2-2. Suggested Section Headings

its final form on each page. Technical illustrators prepare final illustrations from your sketches. IPB writers (catalogers) prepare Illustrated Parts Breakdowns (including helpful exploded views) of equipment. Provisioners prepare lists of support equipment and parts. Photographers photograph the

TECHNICAL WRITING for TECHNICIANS

subject equipment, and may retouch photos to bring out important detail. Reproduction services may range from an office copier to a full-color print-shop.

Many people in these groups are great sources of information, who will give generously of their knowledge, and work with you to produce a complete and accurate manual. You are responsible to provide them with the best possible information, to monitor the course of their progress, and finally to carefully check their completed work for accuracy.

Do not hesitate to return incomplete or incorrect work. In doing so, be tactful. If rework is necessary because you have overlooked something, be sure to take the responsibility. In a small organization, your support may be limited to a layout specialist and access to a draftsperson who will likely work for the engineering department. In this situation, photography and printing may be handled through outside vendors.

OUTLINING

A detailed outline for the new manual must always be prepared before writing can start.

This step consists of expanding on the section headings, providing additional subordinate headings, each with sentences to indicate the information you plan to present at that level.

The detailed outline must be given careful thought to be sure that all factors are included in their logical places, and that nothing has been overlooked. Submittal of a detailed outline may be required by contract, and in any event it must be discussed with the department manager.

Preparation for Writing

To prepare a detailed outline, you will need a "heading schedule" to indicate the subordination of thoughts. Again, the heading schedule may be given in a specification, by a sample manual, or a company style guide. One heading schedule is as follows:

Level 1 Stand-alone, all capital letters without a period

Level 2 Run-in, all capital letters with period, followed by text

Level 3 Run-in, initial capital letters, underlined, with a period, followed by text

Level 4 Run-in, initial capital letters, with period, followed by text

Procedural steps: First level indented, and numbered, with period
Second level indented, with period

A detailed outline of a section using this heading schedule might look like Figure 2-3. Each heading is followed by a sentence describing the intended paragraph content. Planned figures and tables are given titles and sketched roughly.
Figures and tables support the text, and are referenced in the text in numerical order. Figures are placed either in the text, or, with the text, if page-sized. Figures are located in the manual in numerical order immediately after the first text reference. An exception would be a foldout figure placed for convenience at the back of the manual and referenced as FO-1, FO-2 etc. A rule in writing is that a paragraph heading must have two or more subordinate headings, or it must stand alone. This construction is shown in Figure 2-3.

TECHNICAL WRITING for TECHNICIANS

SECTION I
DESCRIPTION

GENERAL
The Clock System, Part Number 1004527, is a solid-state assembly on three circuit boards, which provides system timing to the associated... (This is an introductory paragraph telling what the device is. If short, it can stand without a heading.)

CONTROLS AND INDICATORS

PRINCIPAL PARTS. The control panel shown in Figure 1, illustrates the controls and indicators used to operate the clock system...
CONTROL FUNCTIONS. Control functions shown in Figure 1 are described in Table 1

THEORY OF OPERATION
OVERVIEW. Figure 2 is a block diagram of the clock system. From this figure, crystal-controlled oscillator A provides a 15.575 Kc sine-wave output which is counted down by a factor of 10 in block B to provide the system clock frequency. This signal, now a square wave, is amplified by a power amplifer in block C.

THE OSCILLATOR. The oscillator is a temperature-stabilized Colpitts circuit with a slow-response feedback loop..
Feedback Circuit. The feedback circuit consists of resistors R1, R2, and R3 and capacitors C1, C2, and C3.
Temperature-Stabilizing Circuit. The temperature stabilizing circuit uses two thermistors located...

COUNT-DOWN CIRCUIT. The count-down circuit consists of an eight-bit counter permuted to provide a count of ten...

POWER AMPLIFIER CIRCUIT. The power amplifier consists of two power transistors in a class C push-pull ciruit...

Figure 2-3. Example of a Detailed Outline - Partial

SECTION III
WRITING STYLE

HOW TO WRITE CLEARLY

STYLE. Now that you have a detailed outline and a start on your data base, you can now start writing. It isn't all that hard. Some pointers on writing style should help you get started. Writing style is concerned with how to write sentences so they can be easily read, and convey exactly what you want them to mean to the reader.

Clarity of expression comes with an awareness of certain rules, followed by practice. You may find that some sentences and paragraphs are especially difficult to formulate and may need to be rewritten several times to make them easily understood. Take the necessary time as each situation comes up, don't postpone reworking material.

When you are first faced with describing the subject equipment, consider carefully the answer to this question: "What is it?"— not 'what does it do.' For example, the subject equipment may provide personal transportation...that is what it does. It *is*, however a bus, an automobile, a bicycle, or even a skateboard.

Once you have identified the equipment, then go on to describe what it does, and how it does it. As an example:

```
Kappa is an electromechanical machine for use in
the fast-food industry. In operation, it rapidly
applies a shrink-wrapped closure to the top of a
soft drink cup. Operation is semi-automatic,con-
trolled by solid-state components.
Bi-axial shrink film, held on a 2000-ft roll within
```

TECHNICAL WRITING for TECHNICIANS

```
the machine, is pulled by a transport system to a
position below a circular heating chamber where
it is supported by vacuum. The film is then cut to
a 4-1/2 inch length by a hot wire.
When a cup is raised into the heating chamber, the
cup rim meets the film, then as upward motion of
the cup continues,the film drapes over the cup
rim. Further upward movement lifts a flat disc
within the heating chamber, triggering a heating
system. Heat strikes only the sides of the cup,
shrinking the draped film to create a liquid-
tight seal on the cup.
```

Following are notes on conventions for describing markings on control panels, punctuation, sentence structure, and a list of verbs.

PANEL NOMENCLATURE
When actions are taken at a control panel, the **exact** nomenclature appearing on the panel is used in all cases, never abbreviated. Thus, if an indicator lamp on the panel is marked BLOWER ON, the manual will show: "At the control panel, observe the BLOWER ON lamp." Rarely, a panel name is misspelled. In such a case, use the incorrectly spelled name followed by the correct name in parentheses

PUNCTUATION. Punctuation, when correctly used in a sentence, helps to make the intent of the writer clear. Punctuation includes the comma, colon, semicolon, dash, hyphen, parentheses, period, quotation marks, ellipsis, and slash.

The Comma. Use a comma to separate three or more items or phrases in a series.

Example: The console has a meter panel, control panel, and a power supply.

Writing Style

Use a comma to separate thousands, millions, etc. in digits (do not use commas in serial numbers or part numbers).
Example: The altimeter indicated 31,000 ft.
Use a comma to separate two independent clauses joined by a coordinating conjunction.

Example: Record a one for the next MSB of the digital word, and record a zero for all remaining bits.

The Colon and Semicolon. Use a colon to separate several phrases from an introduction.
Example: The control panel performs three functions: switching input voltages, controlling input power, and monitoring output signals.

Use a semicolon instead of a comma to separate sentence segments which already contain commas.
Example: Designation plates (Figure 7-5, item 2; Figure 7-6, item 3) are secured with pressure-sensitive adhesive.

Use a semicolon to separate two independent clauses that are not joined by a coordinating conjunction.
Example: Record a one for the next MSB of the digital word; record a zero for all remaining bits. (Compare this sentence with the similar sentence using commas.)

Dash. Use a short dash between letters and numbers for separation.
Example: See Figure 1-2.

Hyphen. Use a hyphen to join words serving as a single adjective before a noun.
Example: The air-cooling fan is automatically controlled.

3-3

TECHNICAL WRITING for TECHNICIANS

Example: Air-cooling is used throughout the installation.

Do not hyphenate between a number and a unit of measurement when used as a compound modifier.

Example: One 24 V DC battery is required.

When using a hyphen to mean 'through' or inclusive, insert spaces on each side of the hyphen. This construction may be used in tables.

Example: Parts 1 - 4, not parts 1-4. The expression: Parts 1 through 4 is preferred where space permits.

When assigning a negative value to an electrical quantity, a hyphen may be too short to be readily seen. If necessary, enclose a hyphen in parentheses (-) to avoid error, or better yet, use a computer symbol or hand-draw the negative symbol.

Parentheses. Use parentheses to enclose an abbreviation on the first occurrence of the term.

Example: This documents the RM20AC03 Die Filing Machine (DFM).

Use parentheses to enclose figure and table references in sentences.

Example: The small impeller (Figure 3-2) circulates fluid in the tank.

Period. Use a period to end complete sentences.

Writing Style

Example: See Figure 4-6.

Use a period to indicate a decimal between numbers.

Example: 25.5 V

Use a series of periods to form trailers between titles and page numbers.

Example: 1-3   The Oscillator-Amplifier 135

<u>Quotation Marks.</u> Use quotation marks around spoken comments which are printed.

Example: Miss Jones said, "Go home." Note the position of the period.

<u>Ellipsis.</u> An ellipsis, consisting of three dots, is used to show an omission in quoted material, or an interruption of thought.

Example: The king asked no... treatment... "We must succeed," he said.

<u>Slash.</u> The slash is used to form fractions, or represent the word "per."

Example: 15/16, 45 ft/s

SENTENCE STRUCTURE. Some sentences just seem to ramble, others confuse, and some are beautifully clear and immediately understandable. To help you write clearly, some general rules are given here, followed by a selection of verbs which are favored in technical writing.

TECHNICAL WRITING for TECHNICIANS

<u>Sentences.</u> In the list of rules, each rule is followed by a capital letter, or letters in parentheses which refer to the following examples which illustrate that rule. The examples are headed **BAD**, **GOOD**, and **WHY**.

Contrast the bad example with the good example below it, then find out why the sentences were classified that way.

- Delete unnecessary words. .... (C, E, K, M, S)
  - Use "with" instead of "in conjunction with.".........(A)
  - Use words like "using" or "per" instead of
  "In accordance with."...(C)
  Avoid redundant words ....(B)
  - Avoid pointless repetition of words ....(R)
  - Write to inform, not to impress. ....(B, D, H, T)
  - Consider rewriting any sentence over
  25 words in length, unless it is a list...................(W,X)
- Use basic sentence constructions. ....(L)
  - Use subject/verb/object order whenever possible....(N)
  - Write conditional sentences in **if/then** order. .....(AC)
  - Avoid long introductory phrases or clauses
  before the main clause ....(N)
  - Never write an instructional sentence that
  could cause injury or equipment damage
  if the instruction is followed. ....(A)
  - Delete any sentence that will not help
  the reader do the job ....(F)
  - Start an instruction with a verb. ....(D)
  - Use simple lists. ....(G, J, P)
  - Avoid sentence structures which allow
  alternate meanings. ....(Q,S,U)
- Use active voice whenever possible........................(U)

Writing Style

- Avoid words with multiple meanings..........................(D)
- Avoid judgement words such as "approximate," "fairly," "excessive," "too," or " very" as substitutes for tolerances..............................(A,C)
- Avoid slang expressions. ....(Y)
- Avoid contractions such as "don't," "it's," "we're." ....(AA)
- Use common informal words..........................(C, D, E)
- Write to the audience, not to impress...........(O,W, AB)
- Use telegraphic style, omitting "a," "an," and "the,"only in procedural steps; elsewhere use enough articles to permit ease of reading.............(S)
- Make sure all pronouns have clear antecedents........(Z)
- Use parallel sentence structure..................................(I)

Examples. Here are the examples (A through AC). Study the bad and good examples carefully. Try to discover the reason for the bad/good classification, then read the WHY text.

A. BAD:
> CAUTION: Do not set 50 V DC POWER switch to ON unless guidance amplifiers are disconnected.

GOOD:
> CAUTION: Disconnect guidance amplifiers before setting 50V DC POWER switch to ON.

WHY:
Use positive words to show action to be taken. Negative words can cause confusion.
Never write a sentence so the reader could cause injury or damage equipment by following the instruction in the order written.

TECHNICAL WRITING for TECHNICIANS

B. BAD:
Since change symbols would serve no useful purpose in this extensive revision, none have been used. The manual has been changed to update and expand information as follows:

GOOD:
Since change symbols would not be useful in this extensive revision, none have been used. The manual is changed as follows:

WHY:
Avoid redundant words. Write to inform.

C. BAD:
Prior to unpacking, check the container for any evidence of rough handling during shipment. Damage sustained during shipment should be reported in accordance with current instructions.

GOOD:
Before unpacking, check for evidence of damage. Report any damage using current instructions.

WHY:
Use common informal words. Delete unnecessary words. Use words like "using" or "per" instead of "in accordance with".

D. BAD:
A preliminary visual inspection of the XYZ tester shall be performed to verify that the equipment is complete and that it has not sustained any visible damage.

GOOD:
Inspect the XYZ tester to make sure the equipment is complete and not visibly damaged.

WHY:
Start an instruction with the verb.

Writing Style

E. BAD

During disassembly, note and record the placement and connection of the XYZ kit leadwires and/or component parts. Retain this record to ensure that the repaired kit (or part thereof) can be reinstalled according to the original configuration.

GOOD:
Record the placement of leadwires and parts during disassembly. Use the record for correct reassembly.

WHY:
Eliminate unnecessary words. Use common, informal words.

F. BAD:
This alloy was chosen because of its ability to withstand severe exposure when suitably protected.

REMEDY:
Delete sentence.

WHY: The sentence contradicts itself.

G. BAD:
The section contains cleaning and inspection procedures for the XYZ system and provides troubleshooting and repair procedures for returning defective assemblies to service. Testing procedures and calibration requirements to be used during maintenance are also included.

GOOD:
The section contains procedures to clean, inspect, test, troubleshoot, calibrate, and repair the XYZ system.

WHY:
Use simple lists. Use the active voice. The words " for returning defective assemblies to service" are deceptive.

TECHNICAL WRITING for TECHNICIANS

H. BAD:
The procedures are categorized into active and passive types.

GOOD:
The procedures are either active or passive.

WHY:
The words,"...categorized into" and "types" are important-sounding but carry no information.

I. BAD:
Before working on the unit, connect the chassis ground strap to the work surface and the test equipment must be grounded.

GOOD:
Connect the chassis ground strap to the work surface and ground the test equipment before working on the unit.

WHY:
Use parallel construction.

J. BAD:
Note the figure and index number and the part number assigned.

GOOD:
Note the figure, index, and part numbers.

WHY:
Use simple lists.

K. BAD:
For clarity in describing signal flow, only the sine portion of the synchro circuit will be described here.

Writing Style

GOOD:
For clarity, only the sine portion of the synchro circuit is described.

WHY:
Eliminate unnecessary words. Use the present tense for current conditions.

L. BAD:
Remove the screws of the cover of the drawer.

GOOD:
Remove the drawer cover screws.

WHY:
Use basic sentence construction.

M. BAD:
CAUTION: To avoid a possible faulty connection, do not rewrap a wire that has been previously wrapped on a terminal.

GOOD:
CAUTION: To avoid a faulty connection, do not reuse wire that was previously wrapped.

WHY:
Delete unnecessary words. Do not use the word "possible" unless it adds meaning.

N. BAD:
WARNING: To avoid possibility of electrical shock, prior to removing a unit from the test station, shut down the test station in accordance with paragraph 6-4.

TECHNICAL WRITING for TECHNICIANS

GOOD:
> WARNING: To avoid electrical shock, shut down the test station per paragraph 6-4 before removing the unit from the test station.

WHY:
Avoid long introductory phrases or clauses before the main clause. Use subject/verb/object order whenever possible. The word "per" is acceptable.

O. BAD:
> CAUTION: To prevent accidental short circuits that could damage circuit components, press OFF before removing a circuit card.

GOOD:
> CAUTION: To prevent short circuits, press OFF before removing a circuit card.

WHY:
Delete unnecessary words. The word "accidental" is often used with no meaning. In this instance, there is no need to add, "that could damage circuit components."

P. BAD:
This manual is divided into six sections. Sections I, II, and III provide introduction and descriptions; preparation for use, reshipment and storage; and theory of operation data. Sections IV, V, and VI provide maintenance, parts list, and schematic diagram data.

GOOD:
This manual is divided into six sections:

I	Introduction and Description
II	Preparation for Use, Reshipment and Storage
III	Theory of Operation
IV	Maintenance data
V	Parts list
VI	Schematics

3-12

Writing Style

WHY:
Use simple lists. Do not try to reproduce the entire table of contents in text.

Q. BAD:
Damage to lungs or skin or fire may result if personnel fail to observe safety precautions.

GOOD:
Fire, or damage to lungs or skin can result if personnel fail to observe safety precautions.

WHY:
Avoid sentence structures which allow alternate meanings. It is unlikely that the fire would be damaged.

R. BAD:
These wiring/cabling diagrams consist of those wiring/cabling diagrams for an XYZ special option, which differ from the standard XYZ assembly wiring/cabling diagrams.

GOOD:
These are the wiring/cabling diagrams for the XYZ special option which differ from those of the standard XYZ assembly.

WHY:
Avoid pointless repetition of words.

S. BAD:
Manual contains procedures required to maintain test station.

GOOD:
This manual contains maintenance procedures for the test station.

3-13

TECHNICAL WRITING for TECHNICIANS

WHY:
Use telegraphic style only in procedural steps. Eliminate word "required."

T. BAD:
CAUTION: Use of excessive lubricant will result in the adhesion of unwanted foreign matter to machinery parts, thus reducing the overall effectiveness of the machine.

GOOD:
CAUTION: Too much lubricant will cause dirt to stick to parts, causing wear and degrading performance.

WHY:
Write to inform. Avoid stilted language.

U. BAD:
An alternate plan of pulling the 2000 meter tests from ABC and shooting them at XYZ is being considered.

GOOD:
We are considering moving the 2000 meter tests from ABC to XYZ.

WHY:
Use the active voice, and avoid slang.

V. BAD:
The signal passes zener diode CR1, which limits the maximum voltage, and enters RC delay circuit R2/C3, which causes a delay in signal propagation.

GOOD:
The signal passes zener diode CR1 and enters RC delay circuit R2/C3.

Writing Style

WHY:
The technician knows what zener diodes and RC delay circuits do.

W. BAD: Three card-select signals, uniquely selected by not more than one of the three coax relay driver circuit cards as the destination for each data transfer, are selectively jumpered on each circuit card so that relay driver circuit card A6 is enabled only when the SEL01 (select logic card 01) signal is active; and card A4 is enabled only when the SEL03 signal is active.

REMEDY:
The circuitry must be studied further and its operation reduced to understandable language using short sentences.

WHY:
Consider rewriting any sentence over 25 words in length unless it is a list.

X. BAD:
When a command word for an active OTA01bit is placed subsequently on the output bus from the coax relay interface card, all Coil-Decoder-Disable FF for all switches on all coax relay driver cards that were set by the previous command word, or by a series of command words not containing active OTA01 bit, are immediately cleared, with one exception, by the OTA01 bit to enable the associated coil decoders to energize the selected switch coils.

REMEDY:
Same comment as for W.

WHY:
Same comment as for W.

Y. BAD: Inspect the circuit card for smoked components.

TECHNICAL WRITING for TECHNICIANS

GOOD:
Inspect the circuit card for burned components.

WHY:
Avoid slang.

Z. BAD:
Connect the DVM to the transmitter before it is adjusted.

GOOD:
Connect the DVM to the transmitter before the DVM is adjusted.

.....or... Connect the DVM to the transmitter before the transmitter is adjusted.

WHY:
Make sure all pronouns have clear antecedents. The use of the word "it" confuses the reader.

AA. BAD:
Don't use the DVM in conjunction with other test equipment.

GOOD:
Do not use the DVM with other test equipment.

WHY:
Avoid contractions. Use "with" instead of "in conjunction with."

AB. BAD:
Refer to the appropriate schematic diagrams while reading this theory of operation.

GOOD:
Refer to the schematic while reading this section.

Writing Style

WHY:
Use "schematic" instead of "schematic diagrams." The reader will not refer to inappropriate schematics.

AC. BAD:
Reduce spring tension to approximately 5 lb. if it is too tight.

GOOD:
If the spring tension is not within tolerance, adjust to 5 +/- 1 lb.

WHY:
Write conditional sentences in if/then order. **Do not force the reader to guess what you mean by "too tight"**— provide the tolerance.

VERB SELECTION. The readability of your manual can be improved if you use the capitalized words from this list.

abandon..use DEPART, END, or STOP
accept..use APPROVE, RECEIVE, or SATISFY
accomodate..use AGREE, ALLOW, or CONFORM
accompany..use WITH
accomplish..use COMPLETE or DO
account..use MAKE SURE, REPORT, or TELL
accumulate..use COLLECT or INCREASE
achieve..use ARRIVE AT, COMPLETE, DO, or GET
act..use FUNCTION or WORK
agitate..use STIR or SHAKE
adhere..use FOLLOW or STICK
admit..use ALLOW or ENTER
adopt..use SELECT
advise..use TELL
AFFECT (effect has a different meaning)

TECHNICAL WRITING for TECHNICIANS

affix..use ATTACH
afford..use PROVIDE
aggravate..use INTENSIFY
alleviate..use EASE or LESS
alter..use ADAPT, CHANGE, or MODIFY
anticipate..use EXPECT or PLAN FOR
approach..use NEAR
ascend..use CLIMB or RISE
ascertain..use BE SURE or MAKE SURE
assist..use AID or HELP
associate..use RELATE
assure..use BE SURE, or MAKE SURE
attain..use GET
attempt..use TRY
authorize..use APPROVE
begin..use START
bite..use GRIP, HOLD, or PENETRATE
board..use ENTER, GET ON
bolt..use ATTACH, INSTALL, or SECURE
bring..use DO, OPERATE, or PRODUCE
brush..use APPLY WITH, TOUCH, or WIPE
build..use FORM or MAKE
calculate..use COMPUTE
cease..use END or STOP
cement..use BOND, or GLUE
characterize..use DESCRIBE
chart..use DIAGRAM, or MAP
check..use INSPECT or TEST
choose..use SELECT
clock..use TIME
commence..use START

Writing Style

COMPILE (program organization by computer)
    For other meanings use ASSEMBLE or PREPARE
complement..use ADD TO or COMPLETE
complicate..use MAKE COMPLEX or MAKE DIFFICULT
compose..use GROUP or MAKE
comprise..use INCLUDE or MAKE UP FROM
concern..use ABOUT
conclude..use COMPLETE, END, FINALIZE, FINISH
consist..use CONTAIN or MADE UP OF
constitute..use FORM, MAKE, or PART
construct..use FORM or MAKE
consult..use REFER TO
contract..use REDUCE or SHRINK
contribute..use ADD, AID, or GIVE
correspond..use AGREE, COINCIDE, or CONFORM
counteract..use OPPOSE
create..use CAUSE or MAKE
cutback..use REDUCE
DAMP (to check oscillation of) for wetting, use MOISTEN
deactuate..use DEACTIVATE, NEUTRALIZE, or STOP
decrement..use DECREASE
defer..use DELAY
depict..use SHOW
desire..use REQUEST, REQUIRE, or WILL
detach..use DISCONNECT, DIVIDE, REMOVE, or
        SEPARATE
dictate..use COMMAND or CONTROL
differentiate..use only as a math term
DIG (moving dirt) For mechanical damage, use DENT, GOUGE, or PIT
diminish..use DECREASE or REDUCE
discard..use DISPOSE or REJECT

TECHNICAL WRITING for TECHNICIANS

discern..use OBVIOUS, RECOGNIZE, SEE, or VISIBLE
discolor..use CHANGE COLOR
discontinue..use STOP
discover..use FIND
discuss..use REFER or TELL
dismantle..use DISASSEMBLE, or TAKE APART
dispatch..use SEND
disregard..use IGNORE
distinguish..use DETERMINE, RECOGNIZE, SET APART, or SET OFF
distribute..use DISPENSE, GIVE, PROVIDE or SPREAD
doubt..use SUSPECT
drag..use LAG, PULL, SLOW, or RAIL
draw..use MAKE, MARK, PULL, or STRETCH
duplicate..use COPY
effect..use COMPLETE, DO, or MAKE (See AFFECT))
elapse..use GO BY or PASS
eliminate..use DELETE, OMIT, or REMOVE
emerge..use APPEAR, COME OUT, or ESCAPE
employ..use USE
encase..use ENCLOSE
encompass..use ENCLOSE or INCLUDE
encounter..use MEET
enhance..use IMPROVE
ensure..use BE SURE or MAKE SURE
entrap..use TRAP
examine..use INSPECT, INVESTIGATE, LOOK AT, or TEST
exchange..use INTERCHANGE
exclude..use KEEP OUT, OMIT, or SELECT
execute..use COMPLETE, DO, or PRODUCE
exercise..use OPERATE or USE
exert..use APPLY or USE

Writing Style

exhibit..use DISPLAY or SHOW
expedite..use AID, EASE, or SEND OUT
expel..use EJECT, or EXHAUST
explain..use DESCRIBE, GIVE REASON FOR, or SHOW
express..use MEAN, STATE, or SEND
EXTINGUISH (cause to stop burning)
For lights and lamps, use TURN OFF or GOES OFF
facilitate..use AID, DO, or EASE
fault..use DEFECT, FAIL, or MALFUNCTION
feature..use AS, HAS, LOOK, or SAME
figure..use COMPUTE
flip..use TURN
formulate..use DEFINE, MAKE, MEAN, or WRITE
fracture..use BREAK or CRACK
furnish..use APPLY, GIVE, PROVIDE or SUPPLY
FUSE (melt together)
FUZE (equip a bomb with a fuze)
GAUGE or GAGE
generate..use PRODUCE
govern..use CONTROL, DIRECT, or REGULATE
grab..use CATCH or SEIZE
grasp..use GRIP
halt..use STOP
hammer..use HIT, STRIKE, or TAP
hand..use GIVE
hangar..use MOVE, PARK, or PUT
happen..use OCCUR
hide..use IN BACK OF, or OUT OF VIEW
holdoff..use DELAY
ILLUMINATE (to cast light on) For to produce light, use LIGHT

TECHNICAL WRITING for TECHNICIANS

illustrate..use DESCRIBE or SHOW
immerse..use DIP or SUBMERGE
impair..use DAMAGE or DECREASE
impede..use DELAY, RETARD, or RESTRAIN
impose..use ORDER or PUT ON
impregnate..use CONTAIN, FILL, or SATURATE
impress..use CAVITY, DENT, HOLLOW, or PRESS
incapacitate..use DISABLE
incline..use SLANT or SLOPE
indent.. if metal, use DENT
influence..use AFFECT or CAUSE
initiate..use START
inscribe..use WRITE or MARK
insure..use BE SURE or MAKE SURE
introduce..use ENTER, INJECT, PUT IN, START
    or SUBMIT
involve..use AFFECT, APPLY TO, or INCLUDE
issue..use DISPENSE, FLOW, GIVE, or SPREAD
JUSTIFY (a line of type) if not, use GIVE REASON,
    or MAKE SURE
knot..use ENTANGLE or TIE
log..use RECORD
manipulate..use CONTROL, HANDLE, OPERATE
    or POSITION
manufacture..use MAKE
mar..use GOUGE, SCRATCH, or CAUSE SURFACE
    DAMAGE

need..use REQUIRE
notice..use OBSERVE or SEE, NOTIFY or TELL
nullify..use CANCEL or NEUTRALIZE

Writing Style

obtain..use GET
offer..use PROVIDE or SUBMIT
oil..use APPLY or LUBRICATE
optimize..use MAKE BEST, or MAKE MOST
originate..use CAUSE, COMPLETE or START
overcome..use CANCEL, OVERPOWER, or OVERRIDE
pause..use DELAY, STOP, or WAIT
pay..use LOOK TO, LOOK OUT, or SEE TO
permit..use ALLOW
persist ..use CONTINUE, GO ON, or REMAIN
pertain..use APPLY, REFER, or RELATE
pick ..use SELECT
place..use CLASSIFY, LOCATE, POSITION, or PUT
preclude..use PREVENT
prescribe..use COMMAND or ORDER
present..use FURNISH, GIVE, or SHOW
proceed..use CONTINUE or GO
procure..use GET
prolong..use CONTINUE, EXTEND, or LENGTHEN
PROPAGATE (waveforms etc.) use EMIT, GROW,
    RADIATE, or SPREAD
prove..use ESTABLISH, SHOW, or VERIFY
puddle..use COLLECT
purchase..use GET
reach..use ARRIVE or EXTEND
realize..use DO or KNOW
rebound..use BOUNCE or RECOIL
REFER..to text or tables, but SEE figures
relinquish..use GIVE UP or RELEASE
remark..use COMMENT
remedy..use CORRECT
render..use CONVERT, DO, GIVE, or MAKE

3-23

TECHNICAL WRITING for TECHNICIANS

replenish..use ADD TO, FILL, SERVICE, or SUPPLY
resemble..use IS SIMILAR TO
retain..use KEEP
reveal..use SHOW
save..use KEEP, PRESERVE, or RESERVE
screw..use INSTALL or TURN
scrub..use CLEAN or WIPE
SEE..see figures, REFER to tables and text
seem..use APPEAR
serve..use AID, DO, FUNCTION or WORK
shorten..use REDUCE or CUT
shut..use CLOSE
snatch..use CATCH or JERK
stay..use REMAIN
substitute..use REPLACE
suggest..use MAY, MEAN, or SUBMIT
survey..use SCAN, or SEARCH
sustain..use CONTINUE, HOLD, MAINTAIN, or STAND
tack..use FASTEN, PIN, or STITCH
TERMINATE (to extend to a specific point in a circuit)
        For other meanings use END or STOP
tolerate..use ALLOW, APPROVE, PUT UP WITH,
    or SATISFY
trigger..use SET OFF, START, or TOUCH OFF
understand..use KNOW
utilize.. use USE
watch..use LOOK AT or MONITOR
wedge..use JAM or LODGE

Appendix B: [2],[3],[8],[12],[13],[14],[15],[16]

SECTION IV
TEXT PREPARATION

GENERAL

DRAFT. Draft is the first setting-down of the combined text and art of your manual. It is sometimes referred to as the manuscript, abbreviated ms. Use 8-1/2 x 11 inch paper. Set the spacing between lines of text at space-and-a-half. Use one- and one-half inch spacing for left and one inch for the right margin. Use one inch spacing for the top and bottom of the page. This allows room for comments and changes to be written in clearly. If you are using a PC-driven printer, a tractor-feed with a continuous ribbon of paper is a good choice.

Generally, you will not page-number each page at first, until all illustrations are available and their locations in the manual are determined. It is good practice, however to write page numbers lightly with pencil in the margin of your draft. The draft is single-sided, never backprinted.

When all art is accounted for, each page can then be page-numbered using consecutive numbers, or numbered by section (or chapter), for example 1-1, 1-2, 2-1, etc. One method of page-numbering is to type a master sheet of consecutive numbers, then cut up a copy and paste the page numbers to the completed draft using rubber cement, or removable tape. Use the point of an X-acto knife to handle the tiny numbers.

THE STARTING POINT. You may be inclined to start writing with the general description in Section I, but the most

TECHNICAL WRITING for TECHNICIANS

effective place to start is with the controls and indicators and their functions. Then move on to the specifications and installation.

In writing for the military, study the IPB and its alphanumeric parts index. Then start with disassembly/repair, progressing to cleaning/reassembly. Next prepare the schematic diagrams, followed by operational checkout and troubleshooting. Finally, prepare a general block diagram and a detailed block diagram to support the theory of operation.

Actually, this is a learning experience, and you may find some things operate differently than you first supposed, and that the detailed outline must be amended. If you discover a fact which will affect your estimated hours to complete the manual, bring it to your manager's attention as soon as possible.

Be on the lookout for readily available material for use as illustrations and text...possibly from engineering reports, acceptance tests, or other manuals. In this writing business, "inspired borrowing" is completely acceptable, and will save you much time.

Your department may have "canned" text known as "boilerplate" to be used in certain situations. Read any such material carefully to be certain that it exactly fits your application, or that you can change it to fit.

DO NOT not copy or use any copyrighted material, either text or figures, without the owner's written permission.

Text Preparation

Look for the copyright notice in any outside material you are considering. It is usually found on the back of the title page, and may be a sentence, or merely a circled letter c: © with the owner's name and the date.

There is no prohibition against using general ideas found in copyrighted material and adapting them to your use, but you must not use copyrighted material directly without written permission of the copyright owner.

ILLUSTRATIONS. As you write, visualize the illustrations necessary to support the text. Refer to each illustration in numerical sequence in the text, and place the illustration, with a figure title, on a separate sheet following the first text reference. Roughly sketch illustrations at this point. For some sketching hints, refer to Section V.

If one figure is to occupy more than one page, the titles must be identical, except for a parenthetical suffix such as:
Wiring Diagram for the TDS Transmitter (Sheet 1 of 2)
Wiring Diagram for the TDS Transmitter (Sheet 2 of 2)

Foldout illustrations are generally discouraged because of the cost of reproducing them in the finished manual. If you need two pages, one solution is to plan two opposing pages. This must be indicated by a marginal note to the page layout specialist. The artwork will be split by the gutter (binding), and lines running across one page onto the other must be identified, since accuracy in binding cannot be guaranteed. One method is to use alphabetical identifiers such as A-A, B-B, at the crossover. This is shown in Figure 4-1.

TECHNICAL WRITING for TECHNICIANS

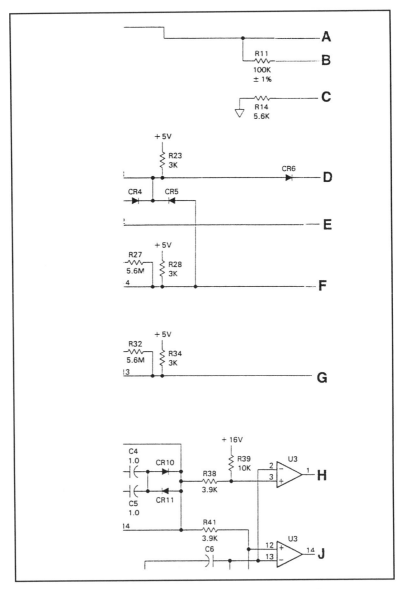

Figure 4-1. Placing a Large Figure on Opposing Pages (Sheet 1 of 2)

Text Preparation

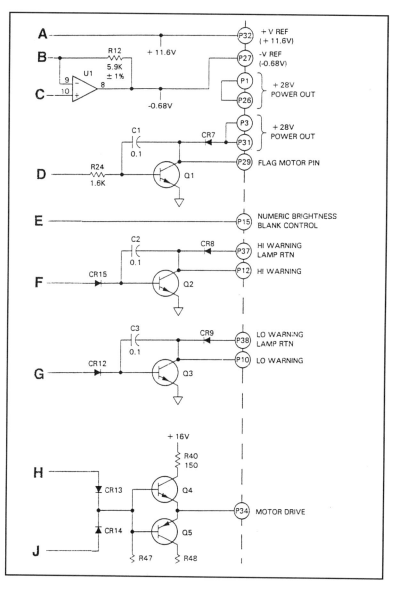

Figure 4-1. Placing a Large Figure on Opposing Pages (Sheet 2 of 2)

TECHNICAL WRITING for TECHNICIANS

TABLES. Tables will take considerable thought to organize so that they will be useful yet reproducible. Unless you have a PC program with which more than 4 or 5 columns can be accomodated, it may be best to treat a complex table as art, in which case you must draw it just as you want it to appear, possibly in pencil, and turn it in as artwork. A sample table is shown in Figure 4-2.

Table references in the text are handled the same as figures, with an in-text reference preceding the table. Notice that the title is at the top of the table.

WARNINGS AND CAUTIONS. As the writer, you are responsible to inform the user of any potential danger to himself or others inherent in the operation or maintenance of the subject hardware. Danger includes death or serious injury. The user is advised by posting a **warning** before directing the user to perform a possibly dangerous step.

Next in importance, is the **caution**, which if ignored, will cause damage to the equipment. Some manuals include a summary of all warnings and cautions in the front matter.

Warnings and cautions must stand alone and cannot be used as procedural steps. Warnings and cautions are commonly displayed prominently in bold type and boxed. A triangle with an enclosed exclamation mark may precede both a warning and a caution.

> ⚠ WARNING: Do not turn on electrical power.

Text Preparation

Warnings and cautions include the following information:

- The specific nature of the hazard
- How to avoid or minimize the hazard
- The location or source of the hazard
- The consequences of disregarding the warning or caution
- Time considerations, when appropriate

A typical warning may look like this:

> WARNING: Shock hazard. The following step applies dangerous potentials up to 7000 volts to exposed terminals and wiring on the oscilloscope chassis. Use extreme caution when working inside the chassis throughout the rest of this procedure.

A caution may look like this:

> CAUTION: Prolonged operation with the INTENSITY control set fully clockwise may burn the screen of the monitor.

RECORD KEEPING

As you start writing, keep a daily log of your work. Date each entry and note progress, problems, and possible solutions. This log will give you material for progress reports, and a history of how you approached the writing task, who you contacted, and their telephone or fax numbers. In addition to this personal log, you may be required to keep a weekly time card on which you list hours spent on each separate task each day.

TECHNICAL WRITING for TECHNICIANS

Item	Name	Part Number	Figure
1	Power Cable	10045263	2-1
2	Signal Cable	10045264	2-1
3	RS 232 Cable	10036589	Not shown

Figure 4-2. Table Example

ESTIMATING

THE PAGE UNIT. In determining the size of a formatted manual, the terms "impression", or "page units" may be used. For example, a sheet of paper that is backprinted, with text or figures on both sides of the sheet, would consist of two page units. A sheet of paper printed on only one side would be one page unit.

DRAFT TO FORMATTED DOCUMENT. On occasion you may be asked to estimate the page units of your manual in its final form while it is still in the draft stage. This requires you to first review the artwork...is it to be reduced and placed "in-text" or will it occupy a full page, or share space on a page with a second figure?

On your double-spaced draft, count the number of words on a couple of pages. Then take a formatted manual written to the same specifications, and do the same. Divide the number of words on a page of the formatted manual by the number of words on your draft page. The resulting figure is the number of draft text pages it will take to make one finished page.

Text Preparation

Multiply this figure by the number of draft pages and add in the artwork. Add front matter and back matter including any index, appendices or addenda which are planned. The total should provide a close estimate.

Be sure, when presenting this information to make note of any "soft" areas of the estimate...those areas that you suspect may change before your manual is complete.

A COMPUTER IN THE HARDWARE

A hardware technical writer will eventually be required to write a manual which describes how to program a computer that is built into the subject hardware. In the past, complex mechanical devices were controlled by gear trains, cams, and cam-operated switches. To change the mode of operation to adapt to changed conditions, gears were shifted and cams were rotated or replaced. Then came the computer. Gear trains, cams and switches have been replaced for timing and sequencing functions.

The computer now drives the system elements (effectors), that do the work, providing the complete timing function through a software program (a schedule of events) carried in the computer.

In some computer-driven equipment, the user can change the program to optimize the device's operation by replacing a computer chip. In others, the user or operator can modify the program by entering the desired information (data) by pressing pushbuttons on a control panel, or by typing commands on a connected PC.

TECHNICAL WRITING for TECHNICIANS

The advent of the built-in computer provides great versatility for electromechanical devices, and requires considerable study by the writer who must provide accurate programming procedures in the manual.

These procedures must be tested on the <u>exact</u> model of the subject device being described in the manual, and must operate flawlessly with the version of software resident in the device. Be aware that a device having built-in Version I software may cause the subject device to behave differently from a device having built-in Version II software.

Perhaps your manual may have to describe operation or programming with both versions. If so, keep the descriptions separate in the text.

Possibly, the latest version (Version II) can be described in the main body of the text, with Version I described in an appendix or in a separate manual supplement. This decision will depend on the number of devices with each version software in the field, and company plans for future support. In this situation, consult with your manager.

Appendix B: [18]

SECTION V
ILLUSTRATIONS

ILLUSTRATIONS

DISCUSSION. Illustrations consist of line art and photographs. When you prepare a sketch, that is the beginning of a line drawing. Photography is used to depict an actual object where a line drawing would not be suitable. Often a photograph of the subject hardware is used as the first figure in a manual to show the user how it actually appears.

Photographs are much more expensive than line drawings, and when used, are commonly black-and-white (B&W), as the cost of printing in color is several times the cost of a black-and-white figure.

TYPES OF DRAWINGS. Artwork other than photographs may be orthographic or three-dimensional, as shown in Figure 5-1. An orthographic figure is two-dimensional, shown flat on the paper (Figure 5-1 A). In the case of a physical object, the front, side, rear, and possibly the bottom are shown as if you were looking straight at these surfaces from a distance. Schematic diagrams, cabling diagrams, and most mechanical drawings are orthographic. To show three sides of a physical object in one figure, a three-dimensional system is required.

True perspective, trimetric, and isometric can all depict an object as having "depth" much as can a 3/4-view photograph. Isometric (Figure 5-1 B) is used in technical writing because it is easier to work with than the other two systems.

TECHNICAL WRITING for TECHNICIANS

Isometric is ideally suited to depicting the "innards" of equipment, by the exploded view technique. In technical writing you will sketch or draw in orthograph or isometric, choosing the best means to support your text. In some instances, you will have on hand an orthographic engineering drawing which must be converted to isometric.

To show the artist (illustrator) what you want, you can take a photograph, or prepare an isometric sketch to accompany the drawings. If preparing a sketch, be especially careful to show all parts in their exact spatial relationship to one another. If it is necessary to rotate a removed part, show it first in its functional position, then use a second view with a note of explanation to the reader.

LINE ART

INTRODUCTION. Line art uses individual lines to form pictures of objects. If the line art is prepared using freehand technique, it is a <u>sketch</u>. If it is done using drafting instruments or Computer-Aided Design (CAD), it is a <u>drawing</u>.

SKETCHING. Sketching is done freehand to depict features of all or part of the subject hardware or of supporting equipment which cannot be clearly described using text alone. Sketching is a valuable skill which will enable you to capture information on the spot using only pencil and paper. The sketch can serve as background information, or as the basis of a finished illustration.

If the sketch will be the basis of a figure, it must be prepared with enough attention to detail so that the artist can readily understand it. A sketch may be orthographic or may also be

Illustrations

drawn to give a perspective, or three-dimensional view as shown in Figure 5-1 B. It is good practice to give the sketch a title, and to note in the margin of the sketch the source of any supporting source information. Sketching takes practice, -- the more you sketch, the better you become at showing detail in its relative size and position. In making an isometric sketch, follow the guidelines given for drawing.

DRAWING. Drawings using drafting instruments such as the T-square, triangle, and scale (ruler) are neater, and may be faster to prepare, or more readily understood than a sketch.

Drawing skill is especially useful if your figure will be used directly, without benefit of an artist, for example to be included with instructions in a quick-fix kit. Drawings used in technical writing are orthographic or isometric, and can be satisfactorily drawn on ordinary copier paper using a mechanical pencil with 0.5mm HB lead. This combination will make a sharp photocopy. The principal tools needed to produce a nice drawing are shown in Figure 1-2.

Over time, certain ways of showing parts on a drawing have become standard. The Code of Lines, shown in abbreviated form in Figure 5-2 is used by all draftsfolk. By applying this code to a mechanical drawing, you can visualize the shape of parts more easily. You will want to follow the code in your own drawing. Electronic technology has its own code of symbols which are described in detail in the ARRL Handbook. Refer to Appendix B[1].

Orthographic Drawings. Orthographic drawings are most often used to show control panels, schematic diagrams, and mechanical parts.

TECHNICAL WRITING for TECHNICIANS

To prepare a sample drawing, square up a sheet of 8-1/2 x 11 inch paper on your drafting board using the T-square, and secure it to the drafting board by its top corners using small squares of removable tape. Now pick a small box-like subject to draw, one that will fit directly on your drafting board. Measure its height, its width and its depth.

You will show the subject in three orthographic views: front, top, and right side.

On your paper, draw a light horizontal center line. Then, using the 30-60° triangle placed on top of the T-square, draw a light vertical center line. See Figure 5-3. These lines give you the center. Divide the subject's height by 2.

Place a mark at this distance on the vertical line on either side of the horizontal line Next, divide the subject's width by 2 and place a mark on the horizontal center line at this distance on either side of the vertical line.

Using the T-square, draw light lines horizontally through the two marks on the vertical center line. With the 30-60 triangle on the T-square, draw light lines vertically through the two marks on the horizontal center line. The intersection of lines should look like the front view of the subject box.

Above the front view, draw two horizontal lines separated by a distance equal to the depth of the subject. These lines, with vertical extensions from the front view, form the top view.

To the right of the front view, draw two vertical parallel lines again separated by the depth. The intersection of these lines and the horizontal extension lines from the front view, give

Illustrations

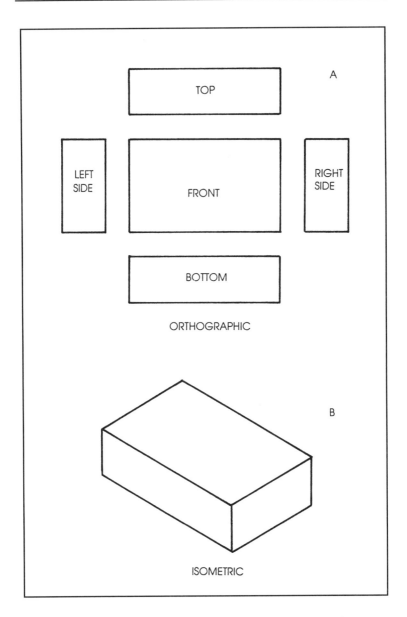

Figure 5-1. Orthographic and Isometric Examples

TECHNICAL WRITING for TECHNICIANS

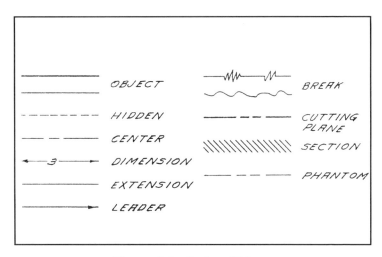

Figure 5-2. Code of Lines

the right side view. To finish the drawing, use an erasing shield and a soft eraser to remove all the construction lines, then darken the lines to bring out the front view, the top view, and the side view.

A natural question to ask is "suppose the subject is much too large to fit on my paper?" The answer is to "scale" it, reducing all dimensions by the same amount, and drawing it in reduced form. This can be figured using a calculator, or with an engineering scale.

Measurements of objects can be made in inches, decimal inches, or in metric. My preference is metric for small items. If you use inches and fractions, the fractions can be converted to decimal values with a calculator by dividing the numerator by the denominator.

Illustrations

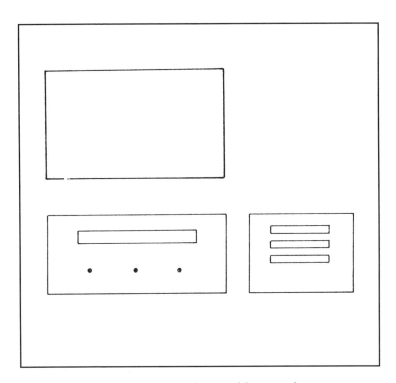

Figure 5-3. An Orthographic Drawing

Thus 1/16 in = 0.0625; 3/32 in = 0.09375 ; and 27/64 in = 0.42187 Now you can easily add them together for a total of 0.57812 in.
This sure beats finding the common denominator!

<u>Isometric Drawings.</u> An isometric drawing, when viewed, creates an impression of reality. In many instances, an isometric drawing is more acceptable than a photograph because it can eliminate clutter. Isometric drawing, a compromise with true perspective, has long been accepted for technical illustration. An isometric drawing uses three major axes 120 degrees apart.

TECHNICAL WRITING for TECHNICIANS

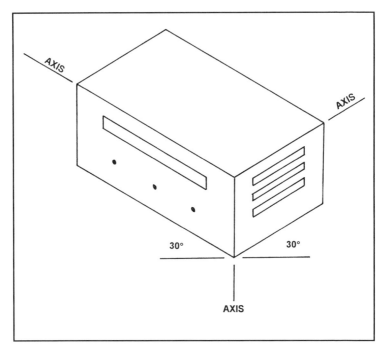

Figure 5-4. An Isometric Drawing

A great advantage of this system is that <u>line lengths along these major axes are proportional to the actual lengths measured on the subject hardware</u>.

The exercise which follows uses the same box used for the orthographic drawing. Begin by squaring a sheet of paper on the drafting board. Find the center of the sheet and make a light mark. Line up the T-square with the center, then using the 30-60° triangle, draw a light line upward to the left from the center at 30 degrees. Flop the triangle over and draw a second light line upward to the right from the center at 30 degrees.

Illustrations

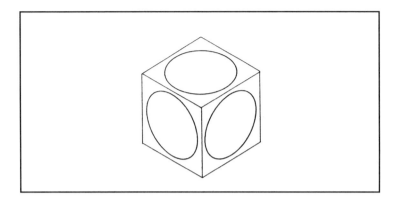

Figure 5-5. A Circle in Isometric

Now slide the T-square down the board and place the triangle upright on the T-square. Line up the vertical edge of the triangle with the center of the sheet and draw a light line from the center downward. You now have all three isometric axes.

Take the measurements from the orthographic drawing. Lay out the length, width, and depth along the axes as shown in Figure 5-4. Next, using T-square and triangle, complete the figure. Notice that you could have shown the front view facing either left or right. This is the basis of isometric drawing.

If on the front of the box there had been a circle, it would appear as an isometric ellipse in this view. This is shown in Figure 5-5. If the circles are small, such as screw heads, they can be sketched, but you will need an isometric template to draw larger ellipses that look "right."

The instructions given for constructing orthographic and isometric drawings are only intended to introduce you to these techniques. You are encouraged to study this subject further.

TECHNICAL WRITING for TECHNICIANS

The best way is to visit the public library and look over their selection of books on engineering drawing (drafting) and technical sketching.

PLANNING. As early as possible, you must decide on the final size of each line drawing and photograph. Art must be planned as either vertical (portrait), or horizontal (landscape). It is further classified as full-page, half-page, or single-column.

Foldouts are expensive to print, but they are especially valuable for electronic and electrical schematic diagrams, and for piping layouts. Some users read the manual, then remove the schematic diagrams for use on the bench or take them into the field. In the case of very large drawings, they can be included at the rear of the manual, possibly in a pocket of the cover.

When you have a sketch that you feel is accurate and complete, <u>make a copy of it for yourself</u>, then prepare a package for the artist (illustrator) consisting of a cover note, the original sketch and all supporting engineering or other information needed to ensure a first quality figure.

The cover note should include:
- Your name, address, and phone number
- Today's date
- The name of the manual in which the figure is to be used
- The figure title, and the section of the manual where it will appear
- The approximate size of the final art: full page, half-page, or dimensions if it is an in-text figure

Illustrations

- Supporting data such as other available documents that use a similar figure, photographs, and engineering drawings you are supplying (List these items.)
- The date you would like the completed figure returned to you

Place a copy of this request in your files with your copy of the sketch.

The artist may prepare the finished art on a drafting board using first pencil then ink, or do the entire illustration on a computer. It is well to check the progress of your work, and ask if there are any supplementary materials the artist needs.

CHECKING ARTWORK. When the figure is completed, you will receive a copy, (or possibly the large up-size board art itself) to check for accuracy. Take time to compare it against your copy of the art sketch. Be sure the art shows accurately what you intended, and that all callouts are spelled correctly.

Tables which have been prepared as art must be checked line-by-line. Notice the size of the callouts...will they take the reduction needed to fit the manual and still remain legible? If changes are in order, pencil them lightly on a tissue overlay, or on a markup copy. Don't ever write directly on the board art if you want the art staff to remain friendly! When the sketch or drawing you submitted is returned to you, put it in a file for future use. It can be changed easier than a copy.

SIZING LINE ART. Artwork as drawn is usually much larger than the space you allocated for it in your manual. This "upsize" art must be reduced to fit. Reduction is figured in percent,

TECHNICAL WRITING for TECHNICIANS

dividing the final dimension by the upsize dimension. Remember, a change in the width will produce a corresponding change in the height.

If you have access to an office copier having reducing capability, the approximate reduction can be entered to obtain reduced copy to place in your draft text. To do this, you may need to make a reduction of a reduction.

Reduction sharpens the lines in the figure, but may cause lines which are too close together to merge into solid black. If you have indicated to the artist the final size you require, the lines and callouts in the figure should reduce cleanly.

TYPE SIZE AND TYPEFACE

Type size is measured in points. 72 points equal one inch. Typefaces belong to families called fonts. Fonts which will concern you are divided into those having typefaces with serifs, and typefaces without serifs, called "sans-serif." Serifs are the fancy curlicues on letters which make them attractive, and lead the eye horizontally for faster reading.

Sans-serif typefaces are plain, and are commonly used for callouts on figures, and in tables as they take reductions well. Sans-serif type leads the eye vertically for rapid scanning of tabular material. Examples of both typefaces are shown below.

Aa Bb Ff   Aa Bb Ff
serif   sans-serif

Illustrations

Type used for manual text may range from 9 to 12 points in size, selected from a serif or sans-serif type font. As a general note, callouts and table entries should not fall below 8 points in their final size, ready for printing.

PHOTOGRAPHS

Photographs for a manual must be planned carefully and their use coordinated with the artist.

<u>Why Photographs?</u> What is the purpose of the photo? Can a line drawing be used instead? If the photo is to be used in the lead-in text, it must be of the exact subject hardware, not a prototype. If used in later text, chances are that it will be used to depict some special feature of the device, and may require callouts.

A photo, to be useful, must be taken using all necessary lighting, preferably in a studio, by an experienced photographer. You must show the photographer by sketches or mark-ups of existing figures, just what features you expect to be captured on film.

You may have the opportunity to be present during a photo shoot of your subject equipment. If it is your first visit, watch carefully and notice the care given to the lighting, the choice of background, and the "posing" of the equipment. Later, you will receive proofs, which you must review with the artist to pick the best one.

The artist must be advised of the final size to be used in your manual so he/she can order a screened print (velox) or screened

TECHNICAL WRITING for TECHNICIANS

negative from the photographer in the exact size. If a computer in conjunction with a scanner is available, the photograph may be "scanned" directly into the computer, callouts added, and a finished art-piece generated on a printer.

In the latest technology, a video pickup is used in the photo studio in place of a conventional camera, and the image is transferred directly to a computer. The final output is on a floppy disc which will go to the artist.

Screening. A photograph or an airbrush illustration are referred to as "continuous tone" since both contain fine gradations ranging from white through shades of gray to black.

The continuous tone illustration cannot be reproduced by the printing process used for manuals unless it is first screened to break the image into tiny dots. These dots are reproduced through the printing process onto the page. As an example, look at a picture in any newspaper with a hand lens and you will see the ink dots which make up the picture. Callouts however, must not be screened or their sharpness will be impaired.

You may sometimes consider using a photo from a printed document. This can be done, but quality suffers since the tiny dots, when printed, tend to spread out on the paper during the printing process, and will never provide as sharp a figure as you can get from an original negative. A new technique using computer-enhancement can improve the quality of a previously printed figure so it can again be satisfactorily printed.

Appendix B:[1],[4],[5],[7],[9]

SECTION VI
REVIEWS

INTRODUCTION

Reviews of a technical manual are performed on one or several copies of the draft manual and again when it is finally formatted and ready for the printer.

IMPORTANT: Never part with your original material. Prepare and give out <u>copies only</u> for review!

IN-HOUSE REVIEWS

Reviews conducted within the company by company personnel are called "in-house" reviews. They are often conducted serially, using one copy of the manual routed in succession to different reviewers. In this case, you must track the manual.

In preparation for review, prepare a sign-off form listing the name of the manual, the name of the first reviewer, the date you forward the manual, and <u>the date you expect it to be returned to you.</u> Include your name and phone number, and address, if appropriate. Make a copy of the sign-off form for your records, then attach the original form to the front of the manual.

As the review progresses, be sure to ask about its progress, showing interest in any developments or setbacks which may impact your schedule.

TECHNICAL WRITING for TECHNICIANS

Reviews are generally arranged through the manager who will look over your work for completeness and adherence to the agreed-upon content. The copy will then be passed to the responsible project engineer for technical comment, then on to marketing. Following each review, you will receive the manual back with pages marked showing needed corrections or additions. If extensive rework is required, make the changes neatly in a second copy of the manual, and route both copies back to the reviewer for signoff.

MILITARY REVIEWS

If you are dealing with the military, review requirements may be given in a supplement to the hardware contract. If the manual is of major interest, there may be several successive reviews at specified points in the generation of the manual.

For example, there may be a review of the detailed outline, a review when the draft is 20% complete, another at 80%, and a final review when the draft manual has been formatted with all final art in place (the finished manual).

Initial reviews may be by mail, or face-to-face with up to 15 reviewers around a table for a week, a "dog and pony show" which will test the writer! A formal review by government representatives must be coordinated by the manager who extends invitations, verifies clearances, finds living accomodations, and schedules in-house support services. Services include a suitable meeting place with a telephone, coordination with engineers to ensure that the subject hardware is up and running and ready for viewing, and secretarial assistance which will be required.

Reviews

The writer prepares one copy of the manual for each attendee plus a couple of extras. When the meeting is in session, each reviewer concentrates on the area of his or her expertise, and shares this knowledge with the others.

The writer must correlate and clarify the comments, entering the group consensus on each point in a copy. Minutes of each session are separately kept by a designated person, usually not the writer, and routed to all attendees. At the end of a session, it is well to review your personal notes against the minutes and quickly resolve any differences.

If steps of operation are described in your manual, the reviewing group may move to a laboratory where the subject hardware is available. Here, each step of your procedure will be tested for accuracy.

If disassembly/assembly is included in your manual, you may be the one to perform the steps exactly as they are written. If all goes well (and the hardware has not been changed), you will receive high marks. Following the review, you will make corrections to the manual, and send a corrected copy to the head of the review board.

On successive reviews of a manual, you may again meet the same reviewers, a serious semi-social affair.

TECHNICAL WRITING for TECHNICIANS

SECTION VII
LAYOUT AND PRINTING

LAYOUT

THE LAYOUT EXPERT. As you prepared the draft manual, you arranged for the figures and tables to follow the first text reference, and may have thought that with a title page, and the addition of figures, the job was completed.

In many companies, when the draft manual has received final approval, it is given to a layout expert who rearranges text, figures and tables for greatest eye appeal, in accordance with the company style guide or other direction. Headers, the space above the text, are prepared for each page to identify the document and section. Footers, below the text, often carry the page number.

Margins may be set for backprinting, wider on the binding edge. Text may be reformatted to two columns, with small figures and tables worked into the text. The text may be converted to a different typeface (font). Assuming the formatting is done on a computer, the table of contents, list of illustrations and list of tables will be automatically (almost magically) prepared.

PROOFREADING. When formatting is completed, you will be given a printout to proofread, and to mark for automatic generation of an index. Do not be overly impressed with the beauty of your manual, instead, pull out your copy of the draft material you gave to the layout expert.

TECHNICAL WRITING for TECHNICIANS

Read and <u>compare</u> <u>every word and sentence of your draft to that of the formatted text.</u> Check artwork and tables for accuracy and legibility. Check the front matter and index (if used) for accuracy. Act <u>immediately</u> to have any discrepancies corrected, and ask for a copy of each corrected page. Check the corrections, and importantly, watch that in making corrections, that the text did not accidently "spill over" onto the next page. Keep a copy of the formatted manual for use as a master in the event someone needs a copy while the originals are at the printer.

IMPORTANT: **The layout effort must not in any way change the technical content of your message**...it is only cosmetic.

You are responsible for the manual, and for catching mistakes which may have crept in along the production path.

PRINTING

When your manual has been formatted and given its final review, a print package must be prepared for the printer. Some companies present the manual to the printer on a floppy disc which contains all text with scanned-in line art and photographs. Other companies use a more detailed print package. In either type submittal, instructions are necessary. Instructions may include identifying and supplying the following items for information:

- The pages of camera-ready (formatted) text with in-text line art and windows for screened photographs.

- Full page line art and photographs.

Layout and Printing

- A make-up sheet which accounts for each page unit of the manual, and tells the printer how to plan his work. If the manual has many oversize photos, a "printer's dummy" may be needed.

The printers dummy is an exact model of the manual as it is to be printed, with all pages and art in place. It is prepared using photocopies of text and artwork.

- Detailed instructions to the printer including a timetable, the number of manuals to be printed, the kind of printing stock and cover stock to be used, and packing and shipping instructions. Generally, manuals should be wrapped and sealed in plastic, and packed in cartons not to exceed 40 pounds.

- Instructions to supply silverprints (bluelines). These are made by the printer on request, from the negatives just before going to press, and represent the last chance to make corrections. Silverprints will show the existence of lines that are too light to print well, or lines that will merge when printed. Silverprints are expensive, but when a large printing is planned, they are worthwhile.

- Directions for return of text, artwork, and printing plates. If the manual is to be reprinted frequently, the printing plates are sometimes left with the printer.

Appendix B: [7]

TECHNICAL WRITING for TECHNICIANS

SECTION VIII
WRITING A SAMPLE MANUAL

Now that you have gotten this far, it is only fair to show you how to put all the information together in an actual manual. The subject hardware is a five-volt power supply with a remote-sensing feature. This power supply is an ArcherKit[R], Catalog Number 28-4015, custom manufactured by Radio Shack, a Division of Tandy Corporation.

This device will provide enough material for each section of a small installation and operation manual. As noted earlier, a manual is usually backprinted, having both left and right pages, and sections starting on a right-hand page. Here, I shall make an exception. <u>This manual will be shown on right-hand pages only.</u> Applicable notes and comments will be given on the facing left pages.

<u>General Information.</u> The subject five-volt power supply provides an adjustable output of 5 Volts +/- 10% (5.5 to 4.5 volts) at a maximum current of 3A at its output terminals. The supply voltage is 105 to 135 V AC. The remote-sensing feature compensates for voltage drops in the leads connecting the supply to the load so that the voltage at the load remains constant. Output current is shown by a meter on the front panel. This power supply has application as a power source for experimental computer circuits using TTL logic.

TECHNICAL WRITING for TECHNICIANS

Comment:

The content of the cover must be determined before starting to write text. These are some of the considerations:

The name of the equipment. This may be a name favored by marketing rather than by engineering, and may include a model number or numbers.

Is a photograph needed? Is there a photo of the **exact** equipment available, or is the subject hardware itself available for photography? A line drawing could be used.

How are the company logo and name to be displayed?

Is color to be used? This will affect photography and costs.

In this manual, the front cover, printed on heavy stock, will bear the name of the equipment, the catalog number, the kind of manual, and the name of the manufacturer, together with a black-and-white photograph of the subject power supply. The back of the front cover is blank. The photograph in reduced form will be again used as Figure 1-1.

The front cover will carry the title. This must be the exact name of the subject equipment. A photograph of the equipment may be used to embellish the cover and to provide easier identification of the subject equipment. If a photo is used, all parts of the subject equipment except cabling must appear. No unrelated items should be visible.

[cover]
Writing a Sample Manual

Installation and Operation

FIVE-VOLT POWER SUPPLY WITH REMOTE SENSING

CATALOG NUMBER 28-4015

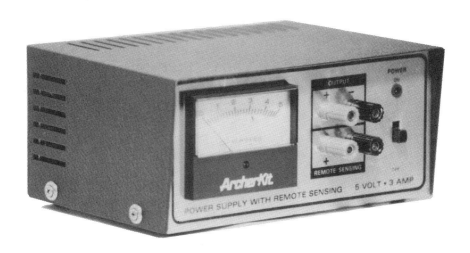

T
C

TECHNICAL WRITING for TECHNICIANS

Comment:

The title page follows the cover, and carries the same information as the cover but may not show the equipment. The title page is printed on the same stock as the text.

[Title Page]
Writing a Sample Manual

Installation and Operation

FIVE-VOLT POWER SUPPLY WITH REMOTE SENSING

CATALOG NUMBER 28-4015

T
C

TECHNICAL WRITING for TECHNICIANS

Comment:

Copyright Notice. The copyright notice, if used, is commonly printed on the back of the title page. The form of the notice as shown is sufficient, but it may be expanded to claim exclusive rights world-wide, and rights in other than printed matter.

Note: The example given here is fictitious.

Writing a Sample Manual

Copyright © 1995
Altex Corporation, Inc.*
San Fernando, CA
U.S.A.

* Altex is a fictitious name used as an example only.

TECHNICAL WRITING for TECHNICIANS

Comment:

The Table of Contents/List of Illustrations/List of Tables, and a listing of Warnings and Cautions generally are included in a manual over 20 pages in length. This information is known as "front matter."

The table of contents will necessarily follow closely the detailed outline which you prepared on receiving this writing assignment. The material for the front matter is unknown at the start of writing, and may best be completed when the manual is nearly finished. Only when the manual is complete, and has been formatted, can page numbers be assigned, and picked up for the front matter, and for an index if provided.

Writing a Sample Manual

CONTENTS

Section		Page
I	Description of the Five-Volt Power Supply	1
	Introduction	1
	Controls and Indicators	1
	Theory of Operation	3
II	Installation	5
	Receiving, Unpacking & Inspection	5
	Planning the Installation	5
	Electrical Connections	5
III	Operation	12
IV	Maintenance & Troubleshooting	13
V	Design Data	19

LIST OF ILLUSTRATIONS

Figure	Title	Page
1-1	The Five-Volt Power Supply	1
1-2	Controls and Indicators	2
1-3	Power Supply Block Diagram	4
2-1	Power and Sensing Leads	8
2-2	Remote Sensing Not Used	9
2-3	Remote Sensing Connections	9
2-4	Series Connections	10
2-5	Split Load Connections	10
2-6	Parallel Connections	11
4-1	Adjustment Location - Cover Removed	14

TECHNICAL WRITING for TECHNICIANS

Comment:

The warnings and cautions are picked up from the text in the order in which they appear. Warnings are listed first, with cautions following

Writing a Sample Manual

LIST OF TABLES

Table	Title	Page
2-1	Power Lead Size	7
4-1	Troubleshooting	15
4-2	Test Equipment	16
4-3	Fuses	18
5-1	Specifications	19

APPLICABLE WARNINGS & CAUTIONS

> WARNING: To avoid electrical shock and possible injury, disconnect the power supply from the 120 VAC source before removing cover

> WARNING: To avoid electrical shock, be especially careful in performing the following procedure. Remove jewelry.

> CAUTION: To avoid damage to the power supply or connected circuits, be sure the POWER on/off switch on the front panel is positioned to off before connecting the power supply to operating power.

> CAUTION: To avoid damage to the power supply, connect it only to an AC 60 Hz 105 to 135 V source.

> CAUTION: To avoid damage to the power supply, do not adjust the foldback current to more than 1.5A.

TECHNICAL WRITING for TECHNICIANS

Comment:

The first page of Section I should give the reader an introduction to the subject hardware, tell what it is, what it does (its purpose), and briefly how it does it. Outstanding features can be listed, and associated equipment noted.

The first page should carry a picture of the subject hardware, possibly a reduced copy of one used on the cover of the manual.

Controls and indicators can be identified as shown here, or if the equipment is simple, an arrow to the control with the control name can be used. If this is done, the use of the controls must be described in the text.

Names that are printed or engraved (placarded) on the equipment are the names to be used in the text. For example, a switch identified with the panel name: BLOWER ON is referred to as the BLOWER ON switch in all references throughout the remainder of the text.

If features of the subject hardware are to be listed, the items of most importance to the user come first. In this example, solid-state circuitry is most important. The metal cabinet is of lesser importance and is placed at the end of the list.

Writing a Sample Manual

SECTION I

DESCRIPTION OF THE FIVE-VOLT POWER SUPPLY

INTRODUCTION. The Five-Volt Power Supply with Remote Sensing, shown in Figure 1-1, is a regulated source of 5-volt electrical power. The power supply is suitable as a source of accurately regulated power for use with electronic circuits which draw up to 3 amperes. Remote sensing compensates for voltage drops in the leads connecting the power supply to the powered device by holding the delivered voltage constant. This action is accomplished by feedback input to the power supply, taken directly from the terminals of the powered device. The output voltage at the powered device varies less than 0.05% from no load to full load. Features include:

- Solid-state circuitry
- Automatic reset
- Fused circuitry
- Diode overload protection
- D'Arsonval ammeter
- Metal cabinet

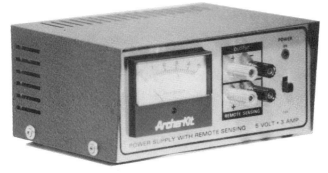

Figure 1-1. The Five-Volt Power Supply

CONTROLS AND INDICATORS
Controls and indicators are shown and described in Figure 1-2.

TECHNICAL WRITING for TECHNICIANS

Comment:

Material for the Controls and Indicators figure can best be prepared by the "hands-on" approach. Try to find an engineering drawing of the face of the control panel, or sketch it carefully. As you view the control panel, notice the placarded indicator lamps...what color(s) do they show when the system is in operation? Is an indicator also a pushbutton? If so, is it a momentary contact type, or alternate-acting?
If fuses are a part of the control panel, check the type and find out the part number. Frequently, a spare fuse will be provided, if so, its location should be noted.

When applying find-numbers to the control panel, start at the most significant feature and proceed either clockwise or counter-clockwise. Do not cross leader lines. For clarity, shaded leader lines are used in Figure 1-2. In the tabular part of the figure, the text is arranged in numerical order.

If there are many items on the panel, and leader lines are crowded, the illustration can be split into two or more parts and separated to provide additional room for leaderlines and item numbers. This technique is especially useful where all electronic components on a circuit board must be identified.

Page 2

Writing a Sample Manual

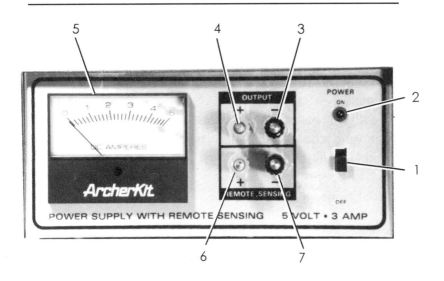

Item	Name	Description	Function
1	POWER ON/OFF	Slide switch	When in the ON position operating power is applied to the power supply. When in the OFF position, power is removed from the power supply.
2	POWER ON/OFF	LED (red)	When lighted, operating power is applied to the power supply.
3	OUTPUT −	Terminal (black)	Negative 5-volt output connection
4	OUTPUT +	Terminal (red)	Positive 5-volt output connection
5	DC AMPERES	Ammeter	Indicates output current in amperes
6	REMOTE SENSING +	Terminal (red)	Positive remote sensing connection
7	REMOTE SENSING −	Terminal (black)	Negative remote sensing connection
8	FUSE	4A	Protection in DC circuit (Fuse at rear, not shown)
9	FUSE	1A	AC input overload protection (Fuse at rear, not shown)

Figure 1-2. Controls and Indicators

8-15

TECHNICAL WRITING for TECHNICIANS

Comment:

The theory of operation should be tailored to the type of manual. An installation and operation manual such as this should explain the general operation of the device to the user. A more detailed theory with accompanying schematic diagrams would be necessary in an overhaul manual. Before writing theory, be sure to confer with your manager on the degree of detail required.

Writing a Sample Manual

THEORY OF OPERATION
The power supply consists of five functional groups shown in the block diagram, Figure 1-3.

AC-to-DC CONVERTER. The AC-to-DC converter is a full-wave transformer-type rectifier which supplies two voltages, nominally 22 V DC and 10 V DC to circuits of the supply. The higher voltage is required by the precision voltage regulator IC. The lower voltage provides the output current, and connects directly to the series-pass regulator. The converter includes the **POWER ON/OFF** switch, the associated LED power on/off indicator, capacitors, a bleeder resistor, and two fuses.

SERIES-PASS REGULATOR. The series-pass regulator consists of a Darlington power transistor in series with the 10 V output of the AC-to-DC converter. This high gain transistor acts as an adjustable power resistor which changes value in response to correction signals from the precision IC regulator.

PRECISION VOLTAGE REGULATOR IC. The precision voltage regulator IC provides a temperature-compensated voltage reference, an operational amplifier, an output driver and a current limiting transistor. The regulator compares the output voltage received from the remote sensing lines to a 5-volt reference. Any difference is amplified and applied to the regulator's output driver. The output driver directly controls the series-pass regulator which changes its resistance to hold the power supply output at exactly 5 volts.

8-17

TECHNICAL WRITING for TECHNICIANS

Comment:

The theory of operation should be tailored to the need of the user. It can be minimal for items such as hand tools, but may be extremely detailed in the case of flight controls for aircraft. Design information which is heavy with mathematics should be referred to in this section, but placed in an appendix unless it is of immediate value to the user.

In this theory of operation, notice that the functional components are first itemized by name, then <u>described in the text in the same order as they were listed, using the same names.</u>

Writing a Sample Manual

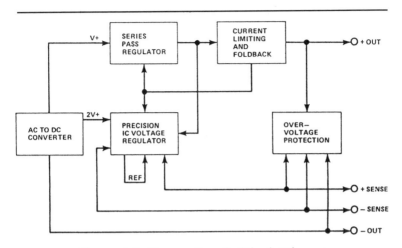

Figure 1-3. Power Supply Block Diagram

CURRENT LIMITING & FOLDBACK CIRCUIT. The current limiting and foldback circuit protects the power supply and the connected device from a current demand in excess of the design limit. The circuit monitors the voltage drop across a resistor which carries the current to the load. When the current rises above a fixed value, a circuit transistor conducts, turning on the current limiting transistor in the precision voltage regulator IC. This action causes the resistance of the series-pass regulator to increase, lowering the output voltage by about 1.3 volts below the 5 volt reference. If the current demand increases, as with a short circuit, the circuit transistor conducts more heavily, and foldback occurs in which the power supply output falls to zero. When the overload is removed, the output voltage returns to its 5-volt level.

OVERVOLTAGE PROTECTION CIRCUIT. The overvoltage protection circuit prevents the output voltage from rising beyond 6.2 volts if a remote sensing line is accidentally disconnected or broken.

TECHNICAL WRITING for TECHNICIANS

In the case of some equipment, the installer may be the manufacturer's representative, or the company's millwright or electrician, not the user. For this reason, installation must be written as accurately as possible, using words and illustrations which cannot be misinterpreted.

If the subject equipment contained a built-in computer, an additional section entitled Programming might be necessary.

Writing a Sample Manual

SECTION II
INSTALLATION

RECEIVING, UNPACKING, AND INSPECTION

The 5-volt power supply was completely tested under operating conditions and thoroughly inspected before shipment. Carefully remove the power supply from the packing. Retain the installation manual. Check the contents against the packing list and inspect the power supply for evidence of damage.

PLANNING THE INSTALLATION

The power supply is intended for indoor use only. It should be placed in a dry location, away from sources of heat. The output power leads and the remote sensing leads should not exceed 6 ft. in length.

ELECTRICAL CONNECTIONS

OPERATING POWER. Operating power connections are made through an attached 6 ft cable located at the rear of the power supply. This device requires an AC 60 Hz, 105 to 135V source of operating power.

> CAUTION: To avoid damage to the power supply, connect it only to an AC 60 Hz, 105 to 135V source.

> CAUTION: To avoid damage to the power supply or connected circuits, be sure the **POWER ON/OFF** switch on the front panel is positioned to **OFF** before connecting the power supply to operating power.

TECHNICAL WRITING for TECHNICIANS

Comments:

Correct output connections are critical to the satisfactory operation of the power supply. Since the writer cannot know all the possible applications of this device, the most common uses must be described, accompanied by wiring diagrams, warnings and cautions. The user should be referred to the device specifications for the maximum load limits.

Writing a Sample Manual

OUTPUT CONNECTIONS. Output connections to the power supply are made at the **OUTPUT** and **REMOTE SENSING** terminals.

NOTE: Remote sensing terminals must always be connected. Refer to the wiring diagrams which follow.

MEASURING THE OUTPUT VOLTAGE. Before connecting the power supply to circuitry, perform the following test using a digital voltmeter (DVM):

1. At the front panel, connect the **REMOTE SENSING** + terminal to the **OUTPUT** + terminal.
2. Connect the **REMOTE SENSING** − terminal to the **OUTPUT** − terminal.
3. Set the voltmeter to the 10 V DC scale.
4. Connect the DVM across the **OUTPUT** + and − terminals.
5. With the **POWER ON/OFF** switch positioned to **OFF**, connect the power supply to operating power.
6. Position the **POWER ON/OFF** switch to **ON**.
7. Measure the output voltage. If it is not 5.0 volts, refer to Section IV for adjustment.
8. Position the **POWER ON/OFF** switch to **OFF**, and disconnect the voltmeter.

WIRE SIZE. The following information will help you to connect the power supply for best operation.

Remote Sensing. The remote sensing feature eliminates the effects of voltage drop through the power leads to the load. At currents up to 2.5 A, a drop of 0.5 V will be compensated for. At currents up to 3.5A, a drop of 0.2 V will be compensated for.

8-23

TECHNICAL WRITING for TECHNICIANS

Comment:

The planning of tables requires considerable effort in order to make them convenient for the user, and to allow them to be reproduced in the manual. As noted in Section IV of the main text, large tables may be spread across opposing pages, or be provided as foldouts. If many foldouts are to be used, consider locating them at the end of the manual.

If the table is a source of critical information for the user, it is well to provide an example of its use.

Writing a Sample Manual

Power Lead Size. Power lead size and length are related in Table 2-1.

Table 2-1. Power Lead Size

Wire Gauge (AWG)	Ohms per ft	Current 2.5A Maximum Length (ft)	Current 3.5A Maximum Length (ft)
22	0.0162	12.5	3.5
20	0.0101	20.0	5.6
18	0.00639	31.5	8.9
16	0.00402	50.0	14.2

Example: Find the maximum length of power leads if No. 18 AWG wire is used at 3.5A. Remember: remote sensing can compensate for 0.2V drop at 3.5 A.

From Table 2-1, No. 18 wire = 0.00639 ohms/ft
Maximum drop per lead = 0.2 V at 3.5A

Find voltage drop per foot: $E = I \times R$

$E = 3.5A \times 0.00639$ ohms

$E = 0.022365$ V per foot

Finally, find the length of conductor which will give a 0.2 V drop:

$$\frac{0.2}{0.022365} = 8.9 \text{ ft answer}$$

8-25

TECHNICAL WRITING for TECHNICIANS

Comment:

Wiring diagrams must be included for all authorized uses of the device. If wire sizes and length of wire runs are a consideration, information must be given. Often, conductors carrying sensitive signals must be shielded, or routed away from other conductors carrying high currents or pulsed currents that could cause interference. Such information should be noted in the text, and on the diagrams.

If different kinds of grounds exist in the equipment, they should be depicted on the wiring diagram, and their uses noted in a "key" placed on the diagram.

Writing a Sample Manual

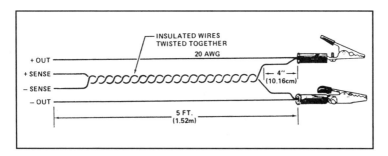

Figure 2-1. Power and Sensing Leads

Connections. Find the correct wire size for power connections from Table 2-1 and the calculation example. Use Number 22 AWG wire for the sensing leads when they are run to the terminals of the powered device. In addition, to avoid electrical interference, twist the sensing leads as shown in Figure 2-1. Be sure to mark the polarity (+/−) of both the power and sensing leads.

WIRING DIAGRAMS. The following wiring diagrams show the correct way to connect the power supply to devices.
Remote Sensing Not Used. If you do not plan to use the remote sensing feature, connect the **REMOTE SENSING** terminals to their respective **OUTPUT** terminals at the power supply as shown in Figure 2-2.
Remote Sensing. To use the remote sensing feature, connect the sensing leads to the terminals of the powered device as shown in Figure 2-3.
Power Supplies in Series. Two or more power supplies may be connected in series to provide a higher voltage at the same rated current. Series connection is shown in Figure 2-4. Note the addition of a diode across each power supply output terminal. The diodes, each rated at 1A, protect the power supplies from reverse-voltage as each power supply is turned on.

TECHNICAL WRITING for TECHNICIANS

Comment:

Twisting the sensing leads reduces the effect of unwanted magnetic fields on the sensitive input circuitry. This same twisting can be used with AC leads to reduce their radiation.

Page 9

Writing a Sample Manual

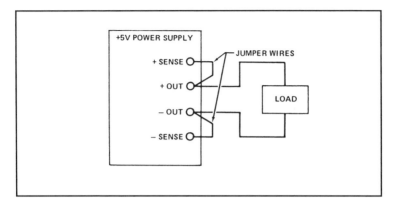

Figure 2-2. Remote Sensing Not Used

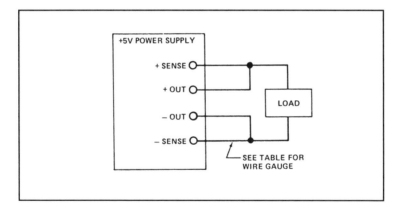

Figure 2-3. Remote Sensing Connections

Power Supplies in Parallel. The first choice, when connecting loads requiring more than 3A, is to split the load as shown in Figure 2-5. If this cannot be done, then two supplies can be connected in parallel as shown in Figure 2-6. Both power supplies must be adjusted for the same voltage, and a 0.1 ohm 5W resistor must be inserted in the positive leg of each power supply connection.

TECHNICAL WRITING for TECHNICIANS

Comment:

In connecting power leads, be sure the standing terminals are clean, and that the lead ends are tinned or provided with crimped or soldered lugs. When operating at full current, use a DVM to check across each output terminal for excessive voltage drop.

Writing a Sample Manual

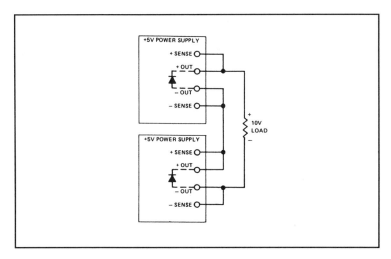

Figure 2-4. Series Connections

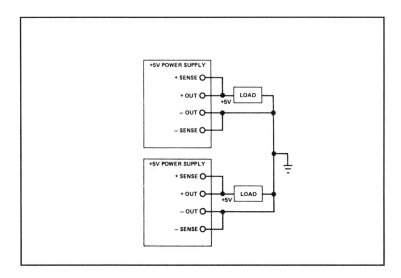

Figure 2-5. Split Load Connections

TECHNICAL WRITING for TECHNICIANS

Comment:

In using power supplies, some general notes apply.

- <u>Before connecting a power supply to a device</u>, turn voltage and amperage controls to minimum, (fully counter-clockwise). Connect a DVM to the output terminals then a p p l y power to the power supply. Set the required voltage u s i n g the DVM, and compare the reading against the panel meter. Turn off the power, remove the DVM, then make connections to the device input connections.

- If both a positive and a negative supply are used, turn on the negative supply first, then the positive. On shutdown, turn off the positive supply first, then turn off the negative supply.

- Do not physically stack power supplies on top of each other so that the cases make electrical contact. Provide some kind of insulation between them.

Writing a Sample Manual

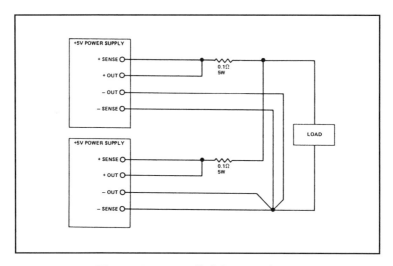

Figure 2-6. Parallel Connections

TECHNICAL WRITING for TECHNICIANS

Comment:

The operation section tells the user how to operate the controls for best results. Warnings and cautions may be needed to prevent personal injury or damage to equipment. Warnings and cautions are always placed <u>immediately before</u> the steps to which they apply and do not replace steps of procedure.

Writing a Sample Manual

SECTION III
OPERATION

Operation of the 5-Volt Power Supply is described in the following general steps:

> CAUTION: To avoid damage to the power supply or connected circuits, be sure the **POWER ON/OFF** switch on the front panel is positioned to **OFF** before connecting the power supply to operating power.

1. At the front panel, position the **POWER ON/OFF** switch to **OFF**.

2. Make power and remote sensing connections as shown in Section II.

3. Carefully check all external wiring.

4. Position the **POWER** switch to **ON**.

5. Read the **DC AMPERES** ammeter on the front panel. If it shows current greater than 3.5A, position the **POWER** switch to **OFF**. Check the load and wiring.

TECHNICAL WRITING for TECHNICIANS

Comment:

Maintenance describes the actions needed to keep the device in efficient operating condition. This includes checking, connections for tightness (power OFF!), examining cables for fraying or damage, and checking system mounting for tightness, especially where vibration is present, and cleaning.

Cleaning instructions must list the types of materials which are approved, for example: "wipe the power supply cabinet and front panel with a soft lintless cloth dampened only with water; do not use alcohol or harsh abrasives."

Often a maintenance schedule is given for guidance. This may include weekly, monthly, and yearly maintenance work plus "When occurring," This last item would cover replacement of failed indicator lamps or fuses.

Certain devices with built-in computers are programmed to conduct extensive self-tests of critical circuits at frequent intervals. If marginal operation or failure is noted during these tests, the user is notified by the lighting of a maintenance indicator, and may be followed by a printout giving helpful details.

Writing a Sample Manual

SECTION IV
MAINTENANCE & TROUBLESHOOTING

MAINTENANCE. Maintenance of the 5-volt power supply is minimal, consisting of inspection for loose terminals or damaged power cord or end connector. The power supply should be occasionally cleaned using a soft lint-free cloth dampened in water. Do not use harsh abrasives.

TROUBLESHOOTING. Figure 4-1 shows the location of adjustments. Table 4-1 lists problems and their remedies.

Cover Removal. To remove the cover, proceed as follows:

> WARNING: To avoid electrical shock and possible injury, disconnect the power supply from the 120 V AC source before removing the cover

1. Turn off electrical power to the power supply by disconnnecting the power plug.

2. Place the power supply on a firm surface.

3. Remove two screws and their cup washers from each side of the power supply.

4. Carefully lift the cover from the power supply.

Cover Installation. Place the cover over the power supply, and install two screws with their cup washers in each side of the cover.

TECHNICAL WRITING for TECHNICIANS

Comment:

When instructing the user to remove a cover, be sure to advise the user of any dangers which may be exposed. This could include high voltages, low voltages at high currents, laser exposure, gear trains, springs, fans, or chemicals.

Warnings and cautions come before the steps to which they apply, but are not themselves a step.

Writing a Sample Manual

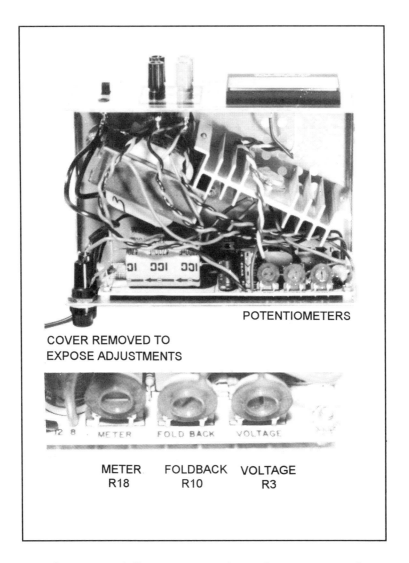

Figure 4-1. Adjustment Location - Cover Removed

TECHNICAL WRITING for TECHNICIANS

Comment:

Troubleshooting provides procedures to isolate and repair equipment failures. Note that the writer must research and list the special tools and test equipment including their part numbers, the procedures to make repairs and the tests, and a means to be certain that a failure has been corrected.

User-replaceable items such as indicator lamps and fuses should be identified in this section. In preparing artwork to show component locations, study engineering drawings and the work of IPB writers (catalogers).

Writing a Sample Manual

Page 15

Table 4-1. Troubleshooting

Trouble	Possible Cause	Remedy
No output voltage	1. Input power connection	Check continuity
	2. POWER switch	Check operation
	3. Fuse F1 or F2	Inspect fuses
	4. Foldback Pot F10	Adjust pot F10 per instructions
	5. Transistor Q3	Replace Q3
	6. Open sense line	Check continuity
	7. Open diode	Check CR1-CR4
Output voltage exceeds 6.5 V	1. Zener VR1	Check VR1
	2. Transistor Q2	Check Q2
Erratic output	1. Load	Check load connections
	2. Remote sense leads	Check sense leads
Output voltage high, but less than 6.5 V	1. Sense leads and connections	Check sense leads and connections
	2. Sense diodes faulty	Check CR3,4,5
	3. R3 out of adjustment	Follow checkout procedure
Output voltage drops when load applied	1. Q1 wiring faulty	Check Q1

8-41

TECHNICAL WRITING for TECHNICIANS

Comment:

A table entitled Special Tools and Test Equipment lists only uncommon tools and the test equipment which is necessary for the tests. Common items such as wrenches and screwdrivers are not included. In this example, no special tools are required, therefore the title has been shortened.

Writing a Sample Manual

CHECKOUT PROCEDURE. The following procedure requires test equipment listed in Table 4-2. Equivalent equipment may be used.

Table 4-2. Test Equipment

Quantity	Item	Identification	Accuracy
1	VOM	Volt/Ohm/Ammeter	3%
4	Resistors	10 Ohm, 10W	10%

Preparation: Connect the four 10 ohm resistors in parallel to provide a 2.5 ohm 40W load.

Procedure. Perform the following steps to adjust the power supply:

1. With the power supply disconnected from 120 V AC, remove the cover from the power supply by following the procedure given in this section.

2. Using a small screwdriver, turn the FOLDBACK potentiometer R10 fully clockwise.

3. Connect the **REMOTE SENSING** + terminal to the **OUTPUT** + terminal.

4. Connect the **REMOTE SENSING** − terminal to the **OUTPUT** − terminal.

5. Adjust the VOM to read 5 V at midscale, and connect it across the **OUTPUT** + and **OUTPUT** − terminals.

8-43

TECHNICAL WRITING for TECHNICIANS

Comment:

The checkout procedure must be absolutely accurate, and easy to follow. Material for the procedure may come from acceptance tests, or from the design engineer. In using in-house procedures, you must ask <u>what kind of test equipment is available to the user</u>, rather than simply listing your production-line checkout tester. This may require consultation with your field-service personnel or the customer.

Writing a Sample Manual

> WARNING: To avoid electrical shock, be especially careful in performing the following procedure. Remove jewelry.

6. Position the **POWER** switch to **OFF**.

7. Connect the power supply to a source of 120 V AC power.

8. Position the **POWER** switch to **ON**.

9. Using a small screwdriver, adjust the VOLTAGE potentiometer R3 so that the VOM reads 5 +/– 0.1 V DC.

10. Position the **POWER** switch to **OFF**.

11. Connect the 2.5 ohm 40W load across the **OUTPUT** + and **OUTPUT** – terminals.

12. Position the **POWER** switch to **ON**.

13. At the VOM, read 5.0 V. If the reading is zero or low, refer to troubleshooting procedures.

14. With a small screwdriver, adjust the **METER** potentiometer R18 so that the panel ammeter reads 2 A.

15. Turn the FOLDBACK potentiometer R10 fully counter-clockwise.

TECHNICAL WRITING for TECHNICIANS

Comment:

When the procedure has been roughed-out it must be tested against the subject equipment for accuracy. As a final check, the procedure should be performed by someone not familiar with the equipment, under the guidance of an experienced technician.

Writing a Sample Manual

16. Short-circuit the output by connecting together the **OUTPUT+** and the **OUTPUT −** terminals.

17. The output voltage and the current shown on the panel ammeter should both go to zero.

> CAUTION: To avoid damage to the power supply do not adjust the foldback current to more than 1.5A.

18. Adjust FOLDBACK potentiometer R10 to bring the panel ammeter reading to 0.5A.

NOTE: If during high-current operation a short circuit across the output terminals causes the panel meter to read zero (0) amperes, remove the short. If the meter still reads zero, the power supply is in a latchup state. While the power supply is still warm, remove power, remove the cover per procedures in this section. Apply power, then reapply the short circuit, and adjust the FOLDBACK potentiometer R10 to produce a slight (200 mA) deflection of the meter needle. Remove power, remove the short circuit, then reinstall the cover.

FUSES. Fuses are located at the rear of the power supply (Figure 1-2). If a fuse requires replacement, be sure that the cause of failure is first corrected.

Table 4-3. Fuses

Fuse	Characteristics	Part No.
F1	1A 250 V Fast-acting	
F24A	4A 32 V Fast-acting	

8-47

TECHNICAL WRITING for TECHNICIANS

Comment:

Design data includes the specifications of the equipment including dimensioned drawings. If optional equipment for use with the subject equipment is available, it should be described here.

Writing a Sample Manual

SECTION V
DESIGN DATA

SPECIFICATIONS. Specifications are listed in Table 5-1.

Table 5-1. Specifications

Item	Specification
Operating power	105 to 135 V AC at 60 Hz
Output voltage	5 V ± 10% internally adjustable
-Voltage variation caused by change in AC input	Less than 0.05% from 105 V to 135 V
-Voltage variation caused by load variations	Less than 0.05 % no load to full load
-Ripple	Less than 1 mV with loads to 3.5 A
Overvoltage protection	Activates at 6.8 V (± 10%)
Current limit	3.5 A (±15%)
Short circuit current	0.5 A (adjustable)
Sense lines	0.5 V per load line to 2.5 A 0.2 V per load line to 3.5 A

8-49

TECHNICAL WRITING for TECHNICIANS

Appendix A   General Information

Resistor Color Code

The color code on tubular resistors with axial leads is read with the color bands turned to the left. Resistors with radial leads use body color, end color, and a colored dot.

First and second color bands from the left have the following values:

Black	0
Brown	1
Red	2
Orange	3
Yellow	4
Green	5
Blue	6
Violet	7
Gray	8
White	9

The third color band is the multiplier. It has the following values:

Black	x	1
Brown	x	10
Red	x	100
Orange	x	1000
Yellow	x	10,000
Green	x	100,000
Blue	x	1000,000
Violet	x	10,000,000
Gray	x	100,000,000
White	x	1,000,000,000

Gold	± 5% of value
Silver	± 10% of value
No color	± 20% of value

Example: You are given a resistor marked 25,000 ohms having a silver tolerance stripe.
The value of the resistor may be 10% higher or 10% lower than its markings indicate: 10% = 0.10

25,000 x 0.10 = 2500, then the value may range from
25,000 − 2500 = 22,500Ω
to 25,000 + 2500 = 27,500Ω

TECHNICAL WRITING for TECHNICIANS

American Wire Gauge to Square Millimeters

AWG	mm²	AWG	mm²	AWG	mm²
30	0.05	16	1.5	2/0	70
28	0.08	14	2.5	3/0	95
26	0.14	12	4	4/0	120
24	0.25	10	6	300 MCM	150
22	0.34	8	10	350 MCM	185
21	0.38	6	16	500 MCM	240
20	0.5	4	25	600 MCM	300
18	0.75	2	35	750 MCM	400
17*	1.0	1	50	1,000 MCM	500

* Not common

Appendix A General Information

Some Constants

Equatorial radius of the earth:　　963.34 statute miles

Acceleration due to gravity
　　at sea level, latitude 45°:　　32.1725 feet/second

Velocity of sound, dry air, 0°C:　　1087.1 feet/second

Velocity of light in vacuum:

$$2.99793 \pm 0.000004 \times 10^{10} \text{ cm./second}$$

Heat of fusion of water at 0°C:　　79.71 cal./gram

Heat of vaporization of water, 100°C:　539.55 cal./gram

One pound of ice absorbs 144 B.T.U in melting

Water weighs approximately 8.336 lb/gallon

Tangent and Sine of 2° = 0.0349

TECHNICAL WRITING for TECHNICIANS

Fahrenheit and Centigrade Temperature Conversions

Water freezes at 32° F (0° C).
At sea level, water boils at 212°F (100°C).

To convert Fahrenheit to Centigrade, use the following formula:

$$°C = \frac{5}{9}(°F - 32)$$

Note that 32 is first subtracted from the Fahrenheit temperature before multiplying by 5/9. The sign of 32 is **minus (−)**.

Example: Convert 70°F to Centigrade.

$$°C = \frac{5}{9}(70 - 32) \; ; \; °C = \frac{5}{9}(38) \; ; \; °C = 21.1$$

Answer: 70° Fahrenheit = 21.1 °C

Appendix A   General Information

Example: Convert −20°F to Centigrade.

$$°C = \frac{5}{9}(-20 - 32)\;;\;°C = \frac{5}{9}(-52)\;;\;°C = -28.89$$

Answer: −20° Fahrenheit = − 28.89 °C

To convert Centigrade to Fahrenheit, use the following formula:

$$°F = \frac{9}{5}°C + 32$$

Note that 32 is **added** to the product of 9/5 x °C

Example: Convert 80° C to °F

$$F° = \frac{9}{5}\,80 + 32\;;\;°F = (1.8 \times 80) + 32\;;\; = 176$$

Answer: 80° Centigrade = 176°F

TECHNICAL WRITING for TECHNICIANS

Example: Convert − 15°C to Fahrenheit.

$$°F = \frac{9}{5}(-15) + 32 \text{ ; or } °F = 1.8 \times (-15) + 32 \text{ ;}$$

°F = −27 + 32 ; °F = + 5

Answer: −15° Centigrade = 5°F

Appendix A General Information

Conversion Factors in Alphabetical Order

TO CONVERT	TO	MULTIPLY BY
Acres (U.S.)	Acres (British)	1.000006
Acres (U.S.)	Square feet	43,560
Acres (US)	Square meters	4,046.8726
Acres (U.S.)	Square miles	0.0015625
Acres (U.S.)	Square yards	4,840
Ampere-hours	Coulombs	3,600
Ampere-hours	Faradays	0.037307
Angstrom Units	Centimeters	1×10^{-8}
Angstrom Units	Inches	3.937×10^{-9}
Angstrom Units	Microns	0.0001
Atmospheres	Cm. of Hg (0°C)	76
Atmospheres	Pounds/Square Inch	14.6960
Barrels (U.S.,liq)	Gal.(U.S., liq.)	31.5
Bars	Dynes/Square Cm	1×10^{6}
B.T.U.	Cal.,gram	252
B.T.U.	Ergs	1.05508×10^{10}
B.T.U.	Foot-pounds	778.184
B.T.U	Joules	1,055
B.T.U/hour	Horsepower	0.000393022
B.T.U./hour	Kilowatts (int.)	0.000293018
Bushels	Cubic Cm.	35,239.3
Bushels	Cubic Inches	1,150.42
Bushels	Gallons (U.S.,dry)	8
Calories, gram	B.T.U.	0.00396832
Calories, gram	Ergs	4.18689×10^{7}
Calories, gram	Foot-pounds	3.08808
Calories, gram	Joules	4.18689
Calories, gram/hr.	Horsepower	1.55964×10^{-6}

TECHNICAL WRITING for TECHNICIANS

TO CONVERT	TO	MULTIPLY BY
Calories, gram/hr.	Kilowatts	1.16279×10^{-6}
Candles/square inch	Lamberts	0.48695
Carats	Milligrams	200
Centimeters	Angstrom Units	1×10^8
Centimeters	Inches	0.3937
Centimeters	Kilometers	1×10^{-5}
Centimeters	Mils	393.7
Cm. Hg (0°C)	Atmospheres	0.0131579
Coulombs	Ampere-hours	0.000277778
Coulombs	Faradays	1.0365×10^{-5}
Cubic Feet	Bushels	0.80356
Cubic Feet	Cubic Inches	1,728
Cubic Feet	Cubic Meters	0.028317016
Cubic Feet	Liters	28.31625
Cubic meters	Cubic Feet	35.314445
Cubits	Feet	1.5
Drams (apoth.)	Grains	60
Drams (apoth.)	Grams	3.8879351
Drams (apoth.)	Ounces	0.125
Dynes	Pounds	2.248089×10^{-6}
Dynes/square cm	Bars	1×10^{-6}
Ergs	B.T.U.	9.47798×10^{-11}
Ergs	Cal.,gram	2.38841×10^{-8}
Ergs	Foot/pounds	7.3756×10^{-8}
Ergs	Joules	1×10^{-7}
Ergs/sec.	Horsepower	1.34102×10^{-10}
Faradays	Ampere-hours	26.8
Faradays	Coulombs	96,480
Fathoms	Feet	6

Appendix A   General Information

TO CONVERT	TO	MULTIPLY BY
Feet	Centimeters	30.48006096
Feet	Miles	0.0001893939
Feet of H2O (60°F)	Atmospheres	0.029469
Feet of H2O (60°F)	lb/square foot	62.364
Feet/hour	Knots	1.64468×10^{-5}
Feet/second	Miles/hour	0.681818
Foot-pounds	B.T.U.	0.00128504
Foot-pounds	Cal.,gram	0.323826
Foot-pounds	Ergs	1.35582×10^{7}
Foot-pounds	Joules	1.35582
Foot-pounds	Kilowatt-hour	3.76617×10^{-7}
Foot-pounds/minute	Horsepower	3.0303×10^{-5}
Furlongs	Feet	660
Furlongs	Meters	201.172
Gallons (dry)	Bushels	0.125
Gallons (dry)	Cubic cm.	4,404.9
Gallons (dry)	Cubic inches	268.8
Gallons (dry)	Gallons (liquid)	1.16364
Gallons (liquid)	Barrels	0.03174603
Gallons (liquid)	Cubic cm.	3,785.434
Gallons (liquid)	Cubic inches	231
Gallons (liquid)	Drams (fluid)	1,024
Gallons (liquid)	Liters	3.785332
Gallons (liquid)	Ounces	128
Gallons/minute	Million gal./day	0.00144
Grains	Carats (metric)	0.32399
Grains	Drams (apoth.)	0.016667
Grains	Ounces (apoth.)	0.00208333
Grams	Carats (metric)	5
Grams	Drams (apoth.)	0.2572059
Grams	Grains	15.432356

TECHNICAL WRITING for TECHNICIANS

TO CONVERT	TO	MULTIPLY BY
Grams	Ounces(apoth.)	0.03215074
Grams/liter	Part/million	1,000
Grams/square cm.	Inches of Hg (32°F)	0.028959
Grams/square cm.	Pounds/square foot	2.04816
Gram-cm.	B.T.U.	9.29472×10^{-8}
Gram-cm.	Cal.,gram	2.34223×10^{-5}
Gram-cm.	Ergs	980.665
Gram-cm.	Foot-pounds	7.2330×10^{-5}
Horsepower	B.T.U./hour	2,544.39
Horsepower	Cal.,gram/minute	10,686.3
Horsepower	Ergs/second	7.45702×10^{9}
Horsepower	Foot-pounds/second	550
Horsepower	Kilowatts	0.745702
Inches	Angstrom Units	2.54000508×10^{8}
Inches	Centimeters	2.54000508
Inches	Mils	1,000
Inches of Hg (32°F)	Atmospheres	0.033421
Inches of Hg (32°F)	Dynes/square cm.	33,864
Inches of Hg (32°F)	Pounds/square foot	70.726
Joules	B.T.U.	0.000947798
Joules	Cal.,gram	0.238841
Joules	Dynes-cm.	1×10^{7}
Joules	Ergs	1×10^{7}
Joules	Foot-pounds	0.73756
Joules	Gram-cm.	10,197.16
Kilograms	Pounds (apoth.)	2.6792285
Kilometers	Feet	3,280.833
Kilometers	Miles	0.6213699
Kilometers	Yards	1,093.611
Kilowatts	BTU./hour	3,412.08

Appendix A   General Information

TO CONVERT	TO	MULTIPLY BY
Kilowatts	Cal.,gram/hour	859,828
Kilowatts	Ergs/second	1×10^{10}
Kilowatts	Foot-pounds/hour	2.65522×10^6
Kilowatts	Horsepower	1.34102
Kilowatts	Kg-meters/hour	3.67098×10^5
Kilowatt-hours	Joules	3.6×10^6
Knots	Kilometers/hour	1.8532486
Knots	Miles (Naut.)/hour	1.00
Knots	Miles/hour	1.15155
Lamberts	Candles/square ft	295.719
Lamberts	Lumens/square foot	929.034
Light years	Kilometers	9.45994×10^{12}
Light years	Miles	5.87812×10^{12}
Liters	Bushels	0.02837819
Liters	Cubic feet	0.03531539
Liters	Gallons(liquid)	0.264178
Liters	Ounces (fluid)	33.8147
Liters	Pints (liquid)	2.11342
Liters	Quarts (liquid)	1.05671
Lumens/square foot	Lamberts	0.00107639
Meters	Angstrom Units	1×10^{10}
Meters	Feet	3.280833
Meters	Inches	39.37
Meters	Yards	1.093611
Meters of Hg (0°C)	Atmospheres	1.31579
Meters of Hg (0°C)	Feet of H2O (60°f)	44.650
Meters of Hg (0°C)	Pounds/square inch	19.3368
Microns	Mils	0.03937

TECHNICAL WRITING for TECHNICIANS

TO CONVERT	TO	MULTIPLY BY
Miles (Naut.)	Feet	6,080.2
Miles (Naut.)	Kilometers	1.8532487
Miles (Naut.)	Miles	1,151553
Miles	Feet	5,280
Miles	Furlongs	8
Miles	Inches	63,360
Miles	Kilometers	1.6093472
Miles (Stat.)	Miles (Naut.)	0.8683925
Miles/hour	Feet/second	1.46667
Millimeters	Inches	0.03937
Million gallons/day	Cubic feet/Second	1.54723
Mils	Feet	8.333×10^{-5}
Newtons	Dynes	100,000
Ounces (fluid)	Cubic cm.	29.5737
Ounces (fluid)	Cubic inches	1.80469
Ounces (fluid)	Drams	8
Ounces (fluid)	Quarts (liquid)	0.03125
Ounces/square inch	Feet of Hg (32°F)	0.0106042
Ounces/square inch	Feet of H_2O (60°F)	0.144314
Pints (liquid)	Cubic inches	28.875
Pints (liquid)	Liters	0.473167
Pints (liquid)	Ounces (fluid)	16
Pounds (apoth.)	Grains	5,760
Pounds (apoth.)	Grams	373.24177
Pounds (apoth.)	Ounces (apoth.)	12
Pounds (apoth.)	Ounces (Avoir.)	13.165714
Pounds (Avoir.)	Grains	7,000
Pounds (Avoir.)	Grams	453.5924277
Pounds (Avoir.)	Ounces (apoth.)	14.58333
Pounds (Avoir.)	Ounces (Avoir.)	16

Appendix A   General Information

TO CONVERT	TO	MULTIPLY BY
Pounds (Avoir.)	Tons (short)	0.0005
Pounds/square foot	Atmospheres	0.000472543
Pounds/square foot	Bars	0.000478803
Pounds/square foot	Cm.Hg (0°C)	0.0359131
Pounds/square foot	Dynes/square cm.	478.803
Pounds/square foot	Feet of Hg (32°F)	0.00117825
Pounds/square foot	Feet of H_2O (60°F)	0.016035
Pounds/square foot	Grams/square cm.	0.488241
Quarts (dry)	Bushels	0.03125
Quarts (dry)	Cubic inches	67.200625
Quarts (dry)	Gallons (liquid)	0.290912
Quarts (dry)	Quarts (liquid)	1.16365
Quarts (liquid)	Cubic inches	57.75
Quarts (liquid)	Drams	256
Quarts (liquid)	Gallons (dry)	0.214842
Quarts (liquid)	Liters	0.946333
Radians	Degrees	57.2958
Radians	Minutes	3,437.75
Reams	Sheets	500
Rods	Feet	16.5
Square cm.	Square inches	0.15499969
Square feet	Acres	2.29568×10^{-5}
Square feet	Square cm.	929.0341
Square feet	Square inches	144
Square inches	Square cm.	6.4516258
Square inches	Square feet	0.00694444
Square kilometers	Acres	247.1044
Square kilometers	Square feet	1.076387×10^7
Square kilometers	Square miles	0.3861006
Square meters	Square feet	10.76387

TECHNICAL WRITING for TECHNICIANS

TO CONVERT	TO	MULTIPLY BY
Square meters	Square yards	1.195985
Square miles	Acres	640
Square miles	Square kilometers	2.589998
Square yards	Square meters	0.8361307
Tons (short)	Kilograms	907.18486
Tons (short)	Pounds (apoth.)	2,430.556
Tons (short)	Pounds (Avoir.)	2,000
Tons (short)	Tons (long)	0.8928571
Tons (short)	Tons(metric)	0.90718486
Watts	B.T.U./hour	3.413
Watts	Cal.,gram/hour	860
Watts	Ergs/second	1×10^7
Watts	Foot-pounds/minute	44.25
Watts	Horsepower	0.001341
Watt-hours	Foot-pounds	2,655.75
Yards	Meters	0.9144018

Some Abbreviations

ANSI	American National Standards Institute
ASCII	American Standard Code for Information Interchange
MCM	One thousand circular mils
MIL-STD	U.S. Military Standard
OSHA	Occupational Safety and Health Administration (U.S. Gov't)

Appendix A General Information

Greek Alphabet

Name	Capital	Lower Case	Name	Capital	Lower Case
Alpha	A	α	Nu	N	ν
Beta	B	β	Xi	Ξ	ξ
Gamma	Γ	γ	Omicron	O	o
Delta	Δ	δ	Pi	Π	π
Epsilon	E	ε	Rho	P	ρ
Zeta	Z	ζ	Sigma	Σ	σ
Eta	H	η	Tau	T	τ
Theta	θ	θ	Upsilon	Y	υ
Iota	I	ι	Phi	Φ	φ
Kappa	K	κ	Chi	X	χ
Lambda	Λ	λ	Psi	Ψ	ψ
Mu	M	μ	Omega	Ω	ω

TECHNICAL WRITING for TECHNICIANS

Mathematical Notation

x	"multiply"
÷	"divide" sometimes a slash as in 3/4 is used
+	"add"
−	"subtract"
√	"square root" If a superscript number is shown to the left of the symbol, it indicates that the root of that number is to be taken of the value under the symbol. For example, a superscript 4 would indicate a fourth root.
<	"is less than"
>	"is greater than"
≥	"is greater than or equal to"
≤	"is less than or equal to"
=	" is equal to"
≡	"is identical with"
≠	"is not equal to"
≅	" is congruent to"
(x, y)	"point whose coordinates are x and y"

On an x-y graph, the horizontal axis is the abscissa, noted as x. The vertical axis is the ordinate, noted as y. Measurements along either axis start at the lower left corner, normally zero. Thus a point on the graph can be identified by coordinates (x,y).

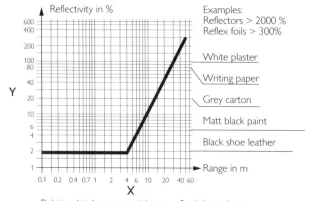

Relationship between minimum reflectivity and range

A-16

Appendix A   General Information

Proofreader's Marks

Mark	Meaning	Mark	Meaning
⊙	Insert period	rom.	Roman type
⋀	Insert comma	caps.	Caps—used in margin
:	Insert colon	≡	Caps—used in text
;	Insert semicolon	c+sc	Caps & small caps—used in margin
?	Insert question mark	=	Caps & small caps—used in text
!	Insert exclamation mark	l.c.	Lowercase—used in margin
=/	Insert hyphen	/	Used in text to show deletion or substitution
∨'	Insert apostrophe		
∨∨	Insert quotation marks	ℓ	Delete
⊥/N	Insert 1-en dash	ℓ̸	Delete and close up
⊥/M	Insert 1-em dash	w.f.	Wrong font
#	Insert space	⌒	Close up
ld>	Insert ( ) points of space	⊐	Move right
shill	Insert shilling	⊏	Move left
∨	Superior	⊓	Move up
∧	Inferior	⊔	Move down
(/)	Parentheses	‖	Align vertically
[/]	Brackets	=	Align horizontally
☐	Indent 1 em	⊐⊏	Center horizontally
☐☐	Indent 2 ems	⊔⊓	Center vertically
¶	Paragraph	eq.#	Equalize space—used in margin
no ¶	No paragraph	∨∨∨	Equalize space—used in text
tr	Transpose¹—used in margin	....	Let it stand—used in text
∽	Transpose²—used in text	stet.	Let it stand—used in margin
sp	Spell out	⊗	Letter(s) not clear
ital	Italic—used in margin	run over	Carry over to next line
___	Italic—used in text	run back	Carry back to preceding line
b.f.	Boldface—used in margin	out, see copy	Something omitted—see copy
∽∽∽	Boldface—used in text	?/?	Question to author to delete³
s.c.	Small caps—used in margin	∧	Caret—General indicator used to mark position of error.
≡≡≡	Small caps—used in text		

[1] In lieu of the traditional mark "tr" used to indicate letter or number transpositions, the striking out of the incorrect letters or numbers and the placement of the correct matter in the margin of the proof is the preferred method of indicating transposition corrections.

[2] Corrections involving more than two characters should be marked by striking out the entire word or number and placing the correct form in the margin. This mark should be reserved to show transposition of words.

[3] The form of any query carried should be such that an answer may be given simply by crossing out the complete query if a negative decision is made or the right-hand (question mark) portion to indicate an affirmative answer.

TECHNICAL WRITING for TECHNICIANS

Sinking and Sourcing Conventions

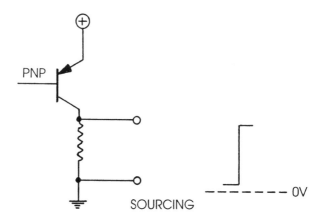

THE OUTPUT SWITCHES THE POSITIVE VOLTAGE TO THE LOAD.
THE LOAD IS CONNECTED BETWEEN OUTPUT AND COMMON.

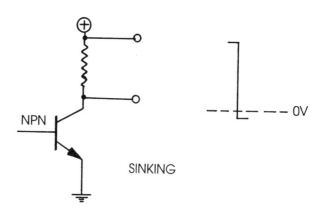

OUTPUT SWITCHES THE COMMON OR NEGATIVE VOLTAGE
TO THE LOAD. THE LOAD IS CONNECTED BETWEEN THE OUTPUT
AND THE POSITIVE SUPPLY.

Appendix B References

[1] ARRL *The ARRL Handbook for Radio Amateurs 1993*. 70th ed.   The American Radio Relay League, 1993

[2] Bates, Jefferson D. *Writing With Precision*. Revised   Acropolis Books Ltd. 1990

[3] Corder, Jim W., and J.J. Ruskiewicz *Handbook of Current English*. 8th ed.   Harper Collins, 1989

[4] Duff, Ronald J. *Applied Sketching and Technical Drawing*   Goodheart-Willcox, 1991

[5] Earle, James H. *Drafting Technology*. 2nd ed. rev.   Addison-Wesley, 1991

[6] Herrington, Donald, *How to Read Schematics* 4th edition   Sams, 1990

[7] International Paper *Pocket Pal, A Graphic Arts Production Handbook*   International Paper Co. 1992

[8] Kirkman, John, *Good Style*. E & FN Spon, an Imprint of Chapman and Hall, London, 1992

[9] Lutz, Ronald J. *Applied Sketching and Technical Drawing*   Goodheart-Willcox, 1991

[10] *Machinery's Handbook*   Industrial Press Inc. 200 Madison Ave.   New York, NY

[11] Mark's *Standard Handbook for Mechanical Engineers*   Baumeister, Avallone, Baumeister   McGraw-Hill Book Company, New York, NY

Appendix B References

[12] Michaelson, H. B. *How to Write & Publish Engineering Papers & Reports.* 3rd ed.　　ORYX Press, 1990

[13] Strunk, William, and White, E.B.　*The Elements of Style.* 3rd ed.　　Macmillan, 1979

[14] University of Chicago　*The Chicago Manual of Style* 13th ed.　　University of Chicago Press 1982

[15] Venolia, Jan *Write Right* Revised Ed.　　Ten Speed Press, 1988

[16] Webster's *Third New International Dictionary,* Unabridged Philip Babcock Gove, editor in chief, G. & C. Merriam Company, Springfield, MA 1976

[17] Webster's *Ninth New Collegiate Dictionary,* G.&C. Merriam Company, Springfield, MA 1984

[18] Wierenga, Moore, Barnes *Procedure Writing Principles & Practices.*　　Battelle Press, Columbus, Ohio, 1993

Index

A

abbreviations	A-14
active voice	3-10
advertising brochures	1-2
alphanumeric parts index	4-2
American Wire Gauge	A-2
Arabic	1-7
ArcherKit	8-1
ARRL Handbook	5-3, B-1
assembly	2-4
autosave	1-8
availability of information	1-2, 2-2

B

block diagram, general, detailed	4-2
boilerplate	4-2

C

camera ready copy	2-7
catalogers	2-1
clear antecedents	3-7, 3-16 (Z)
Code of Lines	5-3
colon and semicolon	3-3
color - indicator lamps	2-3
comma	3-2
command of English	1-1
company benefits	1-4
computer	4-9
constants	A-3
contract reports	1-2
contract writer - See job shop	1-4
contracting officer	2-7
controls and Indicators	2-3, 2-5, 4-2
conversion factors	A-7
copyright notice	1-7, 4-3, 8-6, 8-7
cover stock	7-3
customer	1-7

TECHNICAL WRITING for TECHNICIANS

cutoff or "freeze" date 2-6

D

daily log 4-7
dash 3-3
data base 2-4
deliverable 2-7
deliverable equipment 1-6
detailed outline 2-13
malfunctions 1-6
diskette 1-10
draft preliminary 2-1
drawings 5-3

E

ellipsis 3-5
engineering change notice (ECN, ECO, EO) 2-3
English xi
dictionary 1-8
estimated writer's hours 2-6, 2-7
evening classes 1-2
exploded views 2-11

F

facts 1-6
field service representative 2-7
floppy disc 1-10
foldout illustrations 4-3
fonts 5-12
footer 1-7
formal manual 2-2
freehand drawing (sketching) 5-2
front matter 1-7

G

grade (education level of the user) 2-6
Greek Alphabet A-15

Index

H

halftone figures 1-7
hardware technical manual 1-5
header 1-7
heading schedule 2-13
horizontal (landscape illustration) 5-10
how to unpack 1-6
Human Resources (HR) 1-3
hyphen 3-3

I

Illustrated Parts Breakdown (IPB) 1-2, 2-1
Illustrations 4-3, 5-1
impression 4-8
in-text 4-8
income 1-4
Index 7-1
Indirect hire 1-4
Installation and operatiion manuals 2-1
introduction 1-1
IPB (illustrated parts breakdown) 1-2
isometric drawing 5-1, 5-7, 5-9

J

job description 1-4
job shop (See contract writer) 1-4

K

kinds of manuals 2-1

L

laptop computers 2-2
layout expert 7-1
lead writers (work directors) 2-10
leader lines 8-14
left justified 1-6

TECHNICAL WRITING for TECHNICIANS

list of illustrations 1-6
list of tables. 1-7

M

magnetic fields 1-10
make-up sheet 7-3
manager 2-2
managers xi
manual as a sales tool xi
manuscript 2-1
mathematical notation A-16
metric 5-6
military reviews 6-2

N

non-union occupation 1-1

O

office atmosphere 1-1
operation manual 2-1
ordering replacement parts 1-6
organizing technical material 1-1
orthographic projection 5-1
outside contractor 1-4
outside vendors 2-12
overhaul 2-1

P

page units 4-8
parallel sentence structure 3-7
parentheses 3-4
PC (personal computer) 1-8
perfect binding 1-7
physics. xi
power outage 1-8
preliminary 2-1
price sheets 2-2

Index

printer's dummy 7-3
printed or engraved (placarded) 8-12
printing 7-2
printing stock 7-3
Professional & Technical job classification .. 1-1
project engineer 2-7
proofreading 7-1
Proofreader's Marks A-17
proposal 2-3
prototype 2-3

Q

quotation marks 3-5

R

ragged right 1-7
References B-1
released specifications and drawings 2-6
Resistor Color Code A-1
reviews 6-1, 6-2
right justified margin 1-7
right-hand page 1-7

S

saddle-stitching, 3-ring, perfect binding 1-7
sales engineering 1-2
sample drawing 5-4
sans-serif 5-12
saving text 1-8
scanner 1-8
semicolon 3-3
sentence complexity 2-6
sentences, rules for 3-6
show-through 1-7
sign-off form 6-1
silverprints (bluelines) 7-3
sketching 5-2

TECHNICAL WRITING for TECHNICIANS

slash 3-5
soft-cover books 1-7
specifications, design 2-3
style guide 2-9
support equipment 2-2
supporting services 2-10

T

table of contents 1-7, 8-8, 8-9
tables 4-6
technical background xi
technical editor 2-10
technical illustrators 1-2, 2-11
Temperature Conversions A-4
test specifications 1-2
title page 1-7
top drawing "tree" 2-4
top drawings 2-4
translations 1-6
trimetric projection 5-1
true perspective 5-1
type size 5-12
typefaces 5-12

U

unpacking 1-6, 8-21
unreleased drawings.......................... 2-6

user 2-10

Index

V

verb selection 3-17
version of software 4-10
vertical (portrait) figure 5-10
visualize figures 1-2

W

warnings and cautions 1-7, 4-6
white-collar 1-1
work director 2-10
writer employment 1-2 1-3, 1-4
writer's style guide 2-9
Writing a Sample Manual 8-1
writing assignment 2-2
writing cut-off ("freeze") 2-6
writing file (data base) 2-4

NOTES

TECHNICAL WRITING for TECHNICIANS

ORDER FORM

A great present! A fine addition to a personal library! To order

TECHNICAL WRITING for TECHNICIANS
by W.R. Freeman

Within U.S., please send $19.95 plus $2.50 shipping ($22.45) check or money order for each copy to:

CONTEMAX PUBLISHERS
17815 24th Avenue North
Minneapolis, Minnesota 55447

From:

Name_____

Address_____

City_____ State_____ Zip_____

Please send me_____ I am enclosing $_____
                  number of copies

____Check ____Money Order

Ship to: (If different address)

Name_____

Address_____

City_____ State_____ Zip_____

ORDER FORM

A great present! A fine addition to a personal library! To order

TECHNICAL WRITING for TECHNICIANS
by W.R. Freeman

Within U.S., please send $19.95 plus $2.50 shipping ($22.45) check or money order for each copy to:

CONTEMAX PUBLISHERS
17815 24th Avenue North
Minneapolis, Minnesota 55447

From:

Name_____

Address_____

City_____ State_____ Zip_____

Please send me_____ I am enclosing $_____
                number of copies

____Check ____Money Order

Ship to: (If different address)

Name_____

Address_____

City_____ State_____ Zip_____